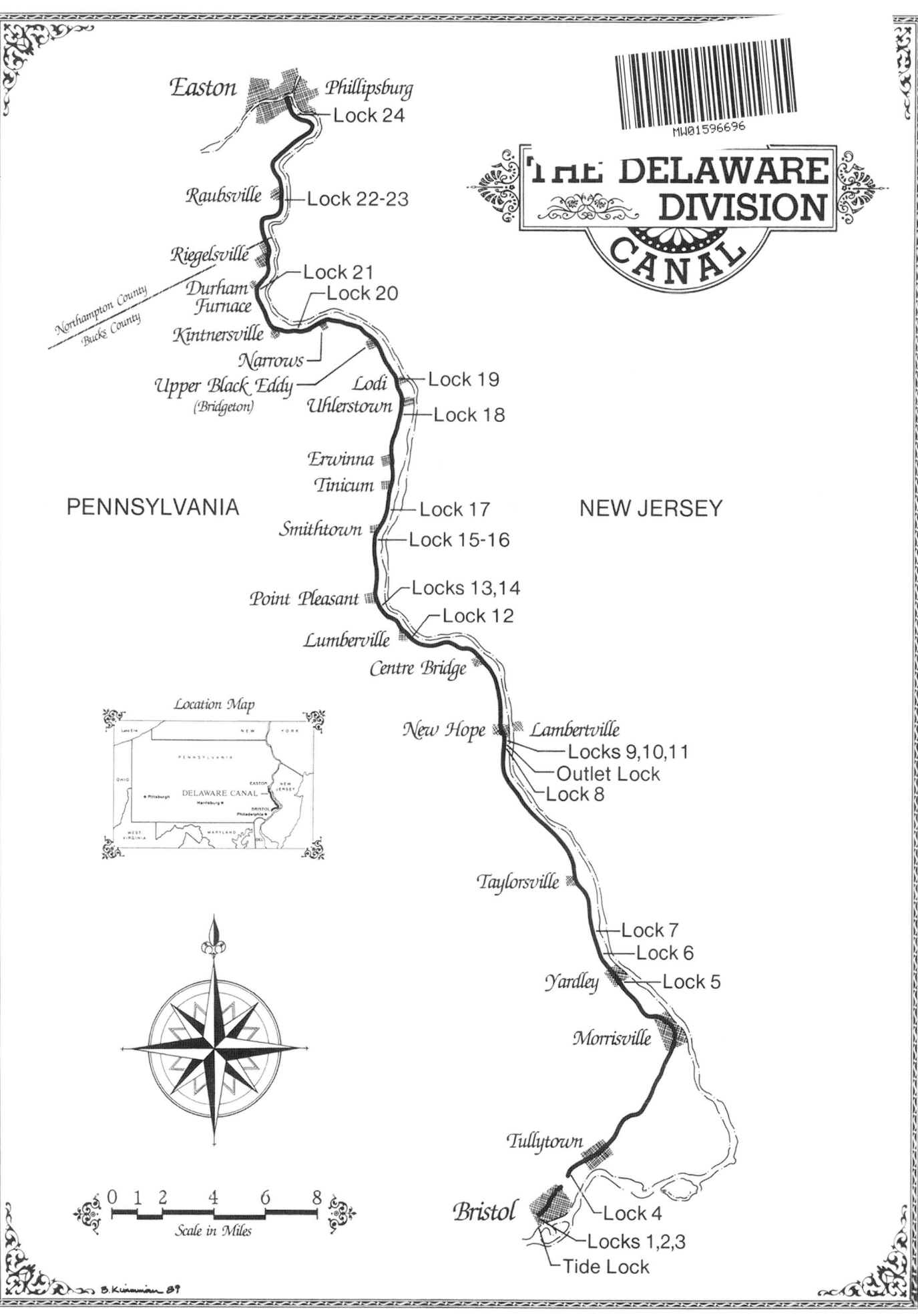

# THE DELAWARE DIVISION CANAL

Easton — Phillipsburg
— Lock 24

Raubsville — Lock 22-23

Riegelsville
Durham Furnace — Lock 21
— Lock 20

Kintnersville

Narrows
Upper Black Eddy (Bridgeton) — Lodi — Lock 19
Uhlerstown — Lock 18

Northampton County
Bucks County

PENNSYLVANIA

Erwinna
Tinicum
— Lock 17
Smithtown — Lock 15-16

Point Pleasant — Locks 13,14
— Lock 12
Lumberville
Centre Bridge

NEW JERSEY

Location Map

New Hope — Lambertville
— Locks 9,10,11
— Outlet Lock
— Lock 8

DELAWARE CANAL

Taylorsville

— Lock 7
— Lock 6
Yardley — Lock 5

Morrisville

Tullytown

Bristol — Lock 4
— Locks 1,2,3
— Tide Lock

0 1 2    4    6    8
Scale in Miles

B. Kinnaman 89

# DELAWARE
# and
# LEHIGH CANALS

*A Pictorial History of the*
*Delaware and Lehigh Canals*
*National Heritage Corridor in Pennsylvania*

## Second Edition

Compiled by Ann Bartholomew

Researched by Lance E. Metz

**Canal History**
**and Technology Press**

National Canal Museum
30 Centre Square
Easton, Pennsylvania, 18042-7743

Copyright © 2005 by Canal History and Technology Press
Published by Canal History and Technology Press
National Canal Museum
30 Centre Square
Easton, Pennsylvania 18042-7743

www.canals.org

*in association with the Smithsonian Institution*

First published in 1989
Second Edition, 2005

ISBN:  978-0-930973-38-4

Library of Congress Control Number:  2005937537

*Printed in the United States of America*

## COVER PICTURES

Front: canalside inn in Bucks County with entrances facing the towpath and the road.
Back: mule tender with his mules at Mauch Chunk;
family traveling south in a stiff boat;
section boat unloading at Leedom's coal yard;
light boat heading north, moving from the Delaware Canal into the Lehigh Navigation.

**Canal History
and Technology Press**

National Canal Museum
30 Centre Square
Easton, Pennsylvania, 18042-7743

# ACKNOWLEDGMENTS

The photographs in *Delaware and Lehigh Canals* are primarily from the collections of the Pennsylvania Canal Society and the National Canal Museum. Additional photographs were made available by the Lehigh County Historical Society and donated by historians Clarence Hendricks, James Lee, John Koehler, and Craig L. Bartholomew.

This book was first published in 1989, shortly after the designation of the Delaware & Lehigh National Heritage Corridor. This second edition, published in 2005, contains a number of additional photographs from the collections of the National Canal Museum.

Staff of the National Canal Museum prepared the first and second editions of *Delaware and Lehigh Canals.* Ann Bartholomew compiled the book, including selecting photos, writing captions, editing and revising text, and designing page layouts. Lance Metz provided research and writings which provided the basis for the text, and assisted in captioning photographs. Mike Knies assisted with selecting photographs, and helped with editing and proofreading. For the second edition, Sarah Hays provided suggestions for revisions to the text, and Martha Capwell Fox combed the archives for photographic images.

Supplemental historical research was provided by Clarence Hendricks, James Lee, Steven Humphrey, and John Koehler. Special insights were provided by former Delaware and Lehigh canal workers Richard Arner, Leon Dreher, and Howard Swope, all now deceased. The maps of the Delaware Canal and the Lehigh Navigation were prepared by Urban Research and Development Corporation, and were based on USGS maps. Wouter de Nie was particularly helpful in correcting and adding to the information that had been presented in the first edition.

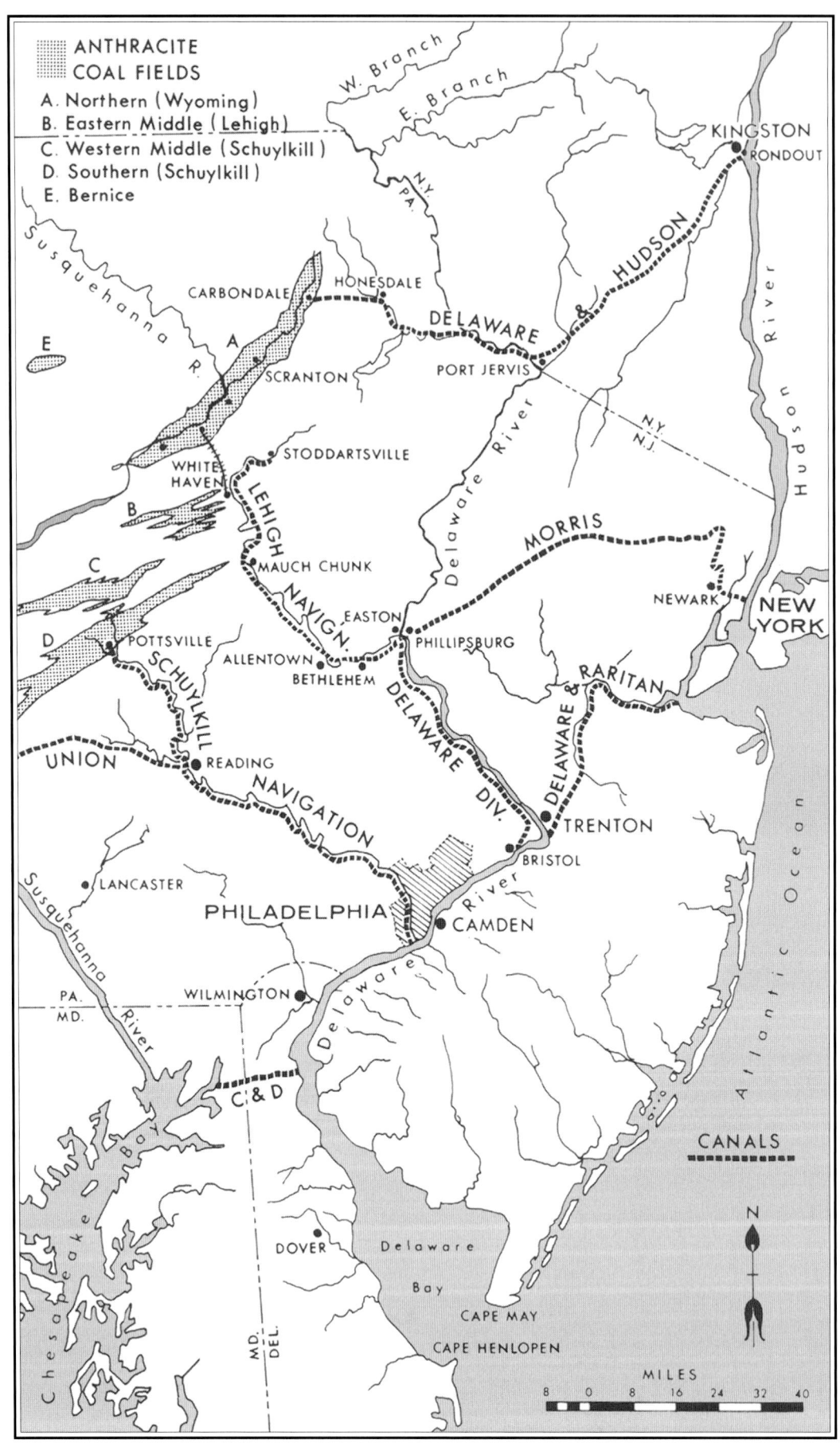

ANTHRACITE
COAL FIELDS
A. Northern (Wyoming)
B. Eastern Middle (Lehigh)
C. Western Middle (Schuylkill)
D. Southern (Schuylkill)
E. Bernice

Coalfields and canals of northeastern Pennsylvania,
showing connections to eastern seaboard cities.

# INTRODUCTION

Passing through a land at times wild, at times placid, aglow with the beauty of nature and befouled by the smokestacks of factories and furnaces, the Lehigh and Delaware canals of the mid-nineteenth century were one of America's major industrial corridors.

The canals carried the raw materials and manufactured products of the Industrial Revolution from the Lehigh Valley to Philadelphia. They interconnected with canal systems in New Jersey that gave access to New York City markets, and they brought New Jersey's ores into the furnaces of the Lehigh Valley.

One, the Lehigh Canal, is the great achievement of two brilliant entrepreneurs of the early nineteenth century, while the Delaware Canal is an example of the Pennsylvania Legislature's program to extend canals throughout the state. Up and down the Delaware and Lehigh rivers the two canals brought opportunities, and a changing way of life. Communities grew, some were born, some later died.

All but abandoned for commercial purposes by the 1930s, the Delaware Canal was recognized by Governor Gifford Pinchot in 1931 as one of Pennsylvania's outstanding scenic areas and designated a state resource, the Roosevelt State Park. As a result, long after its original purpose passed into oblivion, it has been preserved as the most intact of America's towpath canals.

The Lehigh Canal, or more accurately the Lehigh Navigation, as it was a combination of canals and slackwater pools behind dams, has been less well preserved. On the banks of the river, however, there is much evidence of the significance of this waterway. The towns of Jim Thorpe and Catasauqua, and smaller communities such as Weissport, Freemansburg and Glendon owe their existence to the canal.

In the 1980s a growing interest in the reasons for the nation's development and in its heritage — cultural, ethnic, entrepreneurial, industrial — inspired citizens and government to create a National Heritage Corridor along the Lehigh and Delaware canals. The National Canal Museum in Easton interprets the history of these canals and of canals throughout the United States at its museum on Centre Square.

For many years a number of individuals with a keen interest in canals and their history have donated and loaned materials, including thousands of photographs, to the National Canal Museum. This volume has been produced from those materials. Through good years and bad, from the days when the canals were crowded with boats to the days when only a few were seen, these photographs illustrate the people and the sights along the Delaware and Lehigh canals, and commemorate a way of life that has disappeared.

Lehigh Coal and Navigation Company boat 2100 and its crew heading
downstream with a load of anthracite at Lock 16 below Bowmanstown.

# AMERICA'S STONE-COAL TURNPIKE

The story of the Lehigh and Delaware canals begins and ends with coal. Anthracite, or hard coal, from huge reserves in Carbon, Luzerne and Schuylkill counties, was the reason the canals were developed, and was the principal commodity carried on them. The Lehigh Canal was built by a coal company to ship its product; the Delaware Canal was operated by the same company after the state recognized its own inability to do so.

Philadelphia of the late eighteenth and early nineteenth centuries was dependent for its fuel needs on wood, charcoal, and, increasingly, on bituminous coal shipped in from England and Virginia. Both wood and charcoal were becoming scarce and expensive, and little was left of the great forests around Philadelphia. The bituminous—Newcastle and James River coal—was very sooty and not an ideal domestic fuel. Anthracite, on the other hand, burned long and hot, without smoke. Nevertheless, attempts by the owners and lessees of the vast anthracite coalfields of northeastern Pennsylvania to promote their product were mostly unsuccessful. Transporting the anthracite presented extraordinary difficulties that made their coal more expensive than other fuels, and the "stone coal" would neither ignite nor burn successfully in the stoves and fireplaces of the period.

When the coastal blockades by the British during the War of 1812 cut off Philadelphia's bituminous supplies, the owners of the coalfields shipped a few boatloads of anthracite to the city. One was bought by Josiah White and Erskine Hazard, who owned a wire mill and nail works at the Falls of the Schuylkill, north of Philadelphia. They needed a substitute fuel for the scarce charcoal and bituminous coal used in their furnaces and forges, but they were unable even to ignite the anthracite. Finally, in anger or frustration, a workman slammed the door of a furnace and left for the night. Within half an hour, the furnace was glowing with a white heat. The secret of burning anthracite, White and Hazard learned that night, was to use it in a closed stove or furnace and to provide a carefully controlled bottom draft.

White and Hazard were among the most innovative and enterprising men in America. Impressed with this clean-burning, very hot new fuel, they turned their attention to the coal deposits of the Upper Schuylkill Valley, and took a prominent role in the formation of the Schuylkill Navigation Company, which would ship coal down the Schuylkill River to Philadelphia. Following disagreements with other investors, White and Hazard looked into exploiting the coal reserves owned by the Lehigh Coal Mine Company, which had been formed in 1792 to exploit coal deposits near Summit Hill. After determining that they could tame the turbulent Lehigh River, they leased the properties in 1818. The Pennsylvania Legislature granted them virtual ownership of the river, giving

them, in the words of one cynical legislator, "the privilege of ruining themselves" in attempting to make the Lehigh navigable.

During 1818–1820, the first navigation system was constructed. Devised by White, it was a series of dams with a central section that collapsed, creating controlled artificial freshets that floated coal-laden arks downstream. These were known as "bear trap" dams, for no other reason than that when the prototype was being built in the new village of Mauch Chunk, curious onlookers were told that the engineers were constructing a bear trap.

By later standards this was a primitive system as it did not allow boats to return upstream. However, by the end of 1820 more than 365 tons of anthracite had been shipped on arks from Lausanne (above today's Jim Thorpe) 42 miles to Easton, and from there by the Delaware River to Philadelphia. By 1825 the annual tonnage had increased to more than 28,400 tons. Anthracite had become an affordable fuel.

White and Hazard's enterprise was chartered by the Commonwealth of Pennsylvania in 1822 as the Lehigh Coal and Navigation Company.

Although navigation on the Lehigh River had been improved, the Delaware remained in a natural state. White and Hazard were rebuffed in their attempt to build a significantly larger navigation system on the Delaware River. In 1823 they proposed a grandiose scheme to the state, which involved building dams across the river with locks big enough that coastal schooners could be pulled up and down by steam tugboats. Two problems could not be overcome, both of them political. The river was the boundary between two states, so approval of the legislatures of both Pennsylvania and New Jersey would have to be obtained; and lumber interests vehemently opposed the plan, as it would have interfered with their practice of floating large rafts of logs downstream.

White then offered to construct the Delaware Canal at the expense of the Lehigh Coal and Navigation Company in return for toll-free use. The legislature rejected this idea also, and decided in 1827 to build a canal along the Delaware River as part of the state canal system. It was to link Easton with Bristol, a length of 60 miles, and would have 24 lift locks to accommodate the elevation differential of 180 feet.

The state canal was an example of blatant political corruption. Badly designed and poorly built, with narrow, eleven-foot locks that soon proved to be a serious limitation on boats using both canals, it did not even have provision for an adequate supply of water.

The Lehigh Coal and Navigation Company, which would be the principal user of the state canal, dammed the mouth of the Lehigh River and erected abutments to channel Lehigh River water into the Delaware Canal. However when, in 1832, water was first let into the canal it leaked out so rapidly that the entire waterway had to be closed. In desperation, the Pennsylvania Canal Commission asked Josiah White to redesign and rebuild the Delaware Canal, a task he completed in 1834.

The delay caused by the incompetence of the original construction hurt the Lehigh Coal and Navigation Company severely, and the small size of the locks on the Delaware Canal was a constant problem for the company.

During the late 1820s White and Hazard made improvements to the Lehigh Navigation and constructed a gravity railroad to bring coal to the river. A nine-mile gravity railroad from Summit Hill to the loading docks at the new town of Mauch Chunk (Jim Thorpe) was completed in 1827, replacing the rough road from the mines. This railroad was the first one longer than five miles to be built in America. The same year the company began to convert navigation on the river from a downstream-only system to a descending and ascending system.

When it was completed in 1829, the rebuilt Lehigh Navigation employed both slackwater and canal sections capable of allowing the passage of boats carrying almost 200 tons of anthracite. It was 46.2 miles long with a drop of 353 feet, and had locks 22 feet wide by 100 feet long. Ten miles were slackwater pools, 34 miles were canals; there were 8 dams, 5 guard locks (where slackwater ended and dug canal began), 3 guard lifts, and 44 lift locks.

In 1835 the Lehigh Coal and Navigation Company started construction of the Upper Grand Section of the Lehigh Navigation. Intended to extend navigation of the river northward from Mauch Chunk 26 miles to White Haven through the wildest portion of the river, it had to overcome an elevation of 600 feet. To conquer the rapids of the Upper Lehigh, 20 dams and 29 lift locks, some of gigantic size, were built. The highest dam was over 58 feet; the deepest lock could raise and lower boats more than 30 feet. Completed in 1837, the upper section was responsible for the most destructive flood ever to occur in the Lehigh River valley, the flood of 1862, described on page 39.

Instead of extending the navigation system even further north to reach the Susquehanna River near Wilkes-Barre, the Lehigh Coal and Navigation Company made this connection via the Lehigh and Susquehanna Railroad, which became fully functional in 1844–1845 when a series of inclined planes was finished. These planes carried anthracite over the mountains that separated the Susquehanna and Lehigh rivers. America's first wire rope factory was built at Mauch Chunk in 1848 to provide cable for them.

The amount of anthracite shipped from Mauch Chunk increased substantially with the completion of the Delaware Canal and the improvements to the Lehigh Navigation. In 1855, the peak year on the waterway, 1,276,867 tons were shipped. That year, competition from railroads began to erode the amount carried on the waterways, but the total tonnages were still great.

The greatest significance of the canals lies in White and Hazard's desire to increase use of their anthracite beyond existing markets. To this end, they brought a Welsh ironmaster named David Thomas across the Atlantic in 1839 to erect and blow in an iron furnace designed to use anthracite instead of charcoal. The site they selected was Biery's Bridge, later Catasauqua. The first cast of iron was made there on July 4, 1840. From that moment on, anthracite and the canals assumed pivotal importance in the industrial development of the United States.

Using anthracite as fuel in its production, iron for the first time became plentiful and inexpensive. For a period of thirty years, three decades that shaped the future of the Lehigh Valley, anthracite-fueled furnaces throughout the Valley produced greater quantities of iron than any other part of the nation.

With readily available iron, the American Industrial Revolution was soon in full gear. Inventors with innovative mechanical ideas exploited new, limitless opportunities; foundries and factories, large and small, began to produce tools and equipment for a rapidly industrializing nation and for an agricultural community that kept pace with industry in technological improvements. Explosive population growth in the towns took place, starting in the 1840s, as large numbers of people moved from rural areas into industralized areas.

The seminal role of the canals in the ferment of the Industrial Revolution is little appreciated today. Yet it can be traced directly to Josiah White and Erskine Hazard, owners of the Lehigh Coal and Navigation Company, and their quest for expanding markets for their coal. Even after railroads began to take over much of the transportation of raw materials and finished products, the canals remained an integral part of the iron industry's supply and shipping system for many years.

The decline of the canals started with the building of the railroads and continued with the adoption of coke as the primary fuel for smelting iron. As more rail lines were built and, later, as anthracite gave way to coke in the iron and steel industry, and homeowners and factories began to adopt different energy sources for heat and power, less traffic went by canal. By the spring of 1931 only twenty boats remained in operation on the Lehigh and Delaware canals and only 65,000 tons of coal were transported. Later that year, commercial traffic ceased on the Delaware Canal. The Lehigh Navigation stayed in partial operation a few years longer, but repairs necessitated by frequent floods burdened the Lehigh Coal and Navigation Company; eventually, in 1942, almost all repairs were discontinued and the canal was left to nature and a few concerned sportsmen's associations.

# THE MOLLY-POLLY-CHUNKER

As commercial traffic declined, pleasure boating on the canals grew in popularity, particularly on the scenic Delaware Canal.

Interspersed in this collection of photographs are some made by two boating parties in the late nineteenth century, both starting at Bristol and traveling for several days up the Delaware Canal into the Lehigh Navigation. One of these parties is unidentified. The other was a group that included the attorney of the Reading Railroad, Robert W. De Forest, and his wife and daughter; publisher Henry Holt and his daughter; Louise Knox, daughter of the president of Lafayette College; and photographers Walter C. Tuckerman and Louis C. Tiffany, son of jeweler Charles L. Tiffany.

The Molly-Polly-Chunker was the name of their boat. A converted scow, it was home to twelve voyagers, including the crew, from June 15 to 29, 1886, for the trip from Bristol to Mauch Chunk and back — twice the amount of time it took a commercial boat.

The names of the mules, Molly and Polly, were combined with the colloquial term for a boat from Mauch Chunk to give the boat its name. The *Log of the Good Ship Molly-Polly-Chunker, Showing forth the Perilous and Thrilling Adventures of her Company in a Voyage through Strange Countries never before Visited by any Similar Expedition* was published in 1887, with a selection of the photographs taken by Tiffany and Tuckerman.

We shall follow the Molly-Polly-Chunker on its trip from Bristol past Mauch Chunk, from the tide lock on the Delaware River to Easton on the Delaware Canal, and from Easton to the Upper Division of the Lehigh Navigation. However, thanks to the generosity of numerous individuals who have donated a few photographs or entire collections, we shall be able to span the decades in what we see on our trip up the canals.

# THE DELAWARE CANAL

Pennsylvania dismantled its canal system in the mid-nineteenth century after it became clear that railroads were the transportation system of the future. The Delaware Canal was sold to the Sunbury and Erie Railroad Company in April, 1858, and sold again, on July 10 of the same year, to the Delaware Division Canal Company, a subsidiary of the Lehigh Coal and Navigation Company. The name "Delaware Division" has persisted to this day.

A 99-year lease agreement between the Delaware Division Canal Company and the Lehigh Coal and Navigation Company, made in 1866, gave immediate control of the canal to LC&N, which operated it until October 17, 1931. On the same day that the last empty boat went up the canal, the company deeded some 40 miles of canal and right of way, from Raubsville to Yardley, to the Commonwealth of Pennsylvania for use as a public park. The canal was reopened briefly during the summer of 1932 to allow a limited number of boats to carry a few last cargoes of anthracite to coalyards that had encountered problems in securing alternative sources of supply.

Roosevelt State Park (the name was chosen by Governor Gifford Pinchot to commemorate his old friend and fellow preservationist Teddy Roosevelt), was briefly returned to the canal company during the mid-1930s, then given permanently to the Commonwealth in 1940.

The gift of the canal to the people came about because of the sentiment of residents of Bucks and Northampton counties who had grown to love the canal for its quiet beauty, far different from its earlier days as a busy, boisterous navigation corridor. As the Delaware Valley Protective Association, they lobbied and pressured until they prevailed; were it not for them, the company might, and probably would, have sold the right of way to the Commonwealth for use as a roadway. Thanks to them, the Delaware Canal remains today the most intact of America's towpath canals. The Friends of the Delaware Canal, founded in 1982, is a group of volunteers that is continuing the tradition of caring for and about the canal.

Designation as a National Historic Landmark came to the Delaware Canal in 1978. In 1989, the name was changed from Roosevelt State Park to The Delaware Canal State Park. Also in 1989, the Delaware and Lehigh Navigation Canal National Heritage Corridor was created by Congress. This designation allows the National Park Service to assist state and local governments and agencies in preserving and interpreting the canal, and will ensure its preservation for generations to come.

The Swope family taking a load of anthracite south through Bucks County.

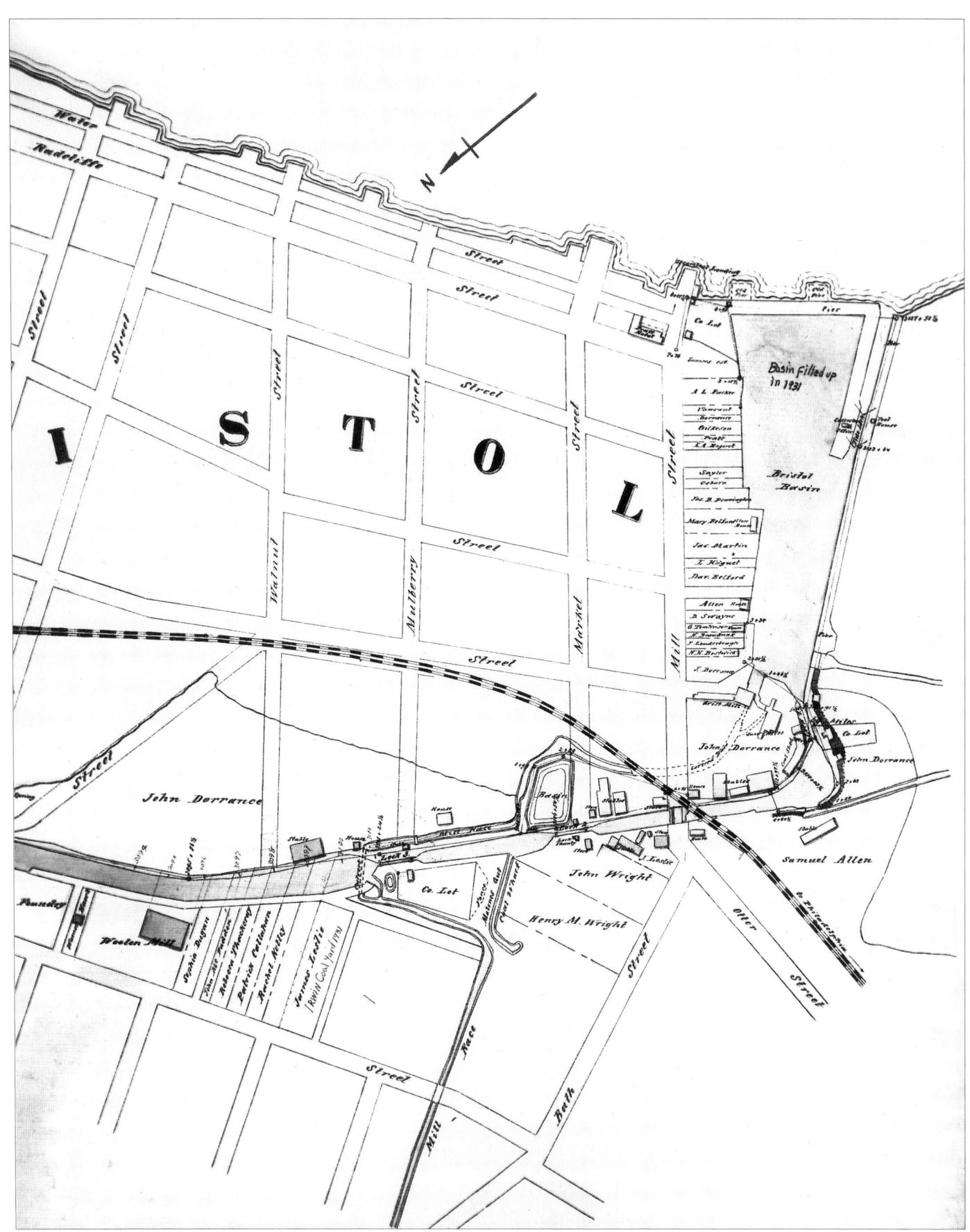

Bristol, on the Delaware River seventeen miles from Philadelphia, was the terminus of the Delaware Canal. From there, coal-laden boats were towed to Philadelphia.

(above) Pleasure boating on the Delaware Canal was a popular pastime in the late nineteenth century. This unidentified group is loading supplies at the tide lock at Bristol for several days of boating up the canal on a custom-built boat drawn by one mule. (right) The passenger boat Burlington, seen here at Bristol, which operated on the Delaware River. (below) Up and down tows in the river at Bristol in 1909. Boats were lashed together and towed by tugboat, the boatmen staying with their boats for the entire trip.

*(top)* Toll collector and toll house at the tide lock. Tolls were collected at Bristol, New Hope, and Mauch Chunk. In the background are some of the industrial buildings on the edge of the canal basin. *(left, above)* Postcard view of the Delaware River, seen from the tide lock. *(left, below)* Boats tied up at the slip outside the tide lock. Sterns of boats show the owner's name. Most of the boats using the Delaware Canal were owned by the Lehigh Coal and Navigation Company, but private boats hauled a wide variety of goods. Generally, loaded boats were coming down the canal and light boats were returning north.

(above) Looking west into the canal basin at Bristol on October 1, 1898, with two loaded private boats on the left. The mill in the background is one of several businesses and warehouses on the north side of the basin. (left) Coal and lumber yard on the north side of the Bristol basin, also on October 1, 1898. Bristol had the largest concentration of coal yards, industries and businesses directly associated with the canal of any community along the Delaware Canal.

(*above*) Unloading an uncoupled "section boat" at Leedom's yard in Bristol in 1931. The rear section has been unloaded first. Only 65,000 tons were carried on the Delaware Canal in 1931, a reduction from a peak of 1,276,000 tons in 1855. The canal closed to commercial traffic in 1931, but reopened briefly in 1932 because Leedom's and other coal yards had not made arrangements to have coal trucked in.

(*right*) An uncoupled section boat, showing the levers used to operate the coupling pins. Most of the boats on the Lehigh and Delaware canals were section boats as they were more maneuverable than stiff boats and could be loaded and unloaded more easily.

*(above)* Locktender's house, basin and overflow lock above Lock 1, at the northern end of the Bristol basin. *(below)* Lock 2 at Bristol. The "doghouse" to the left of the bridge protected the gears used for opening and closing the lock gates.

*(above)* Lock 2 on August 28, 1931. *(below)* Lock 3, looking east.

(above) Private boat, loaded, at Bristol. Private boats had a higher bow and stern than company boats, and had peaked hatch covers. (below) Clearance under the footbridge at locks was a problem only for light boats, which sat high in the water.

(above) Entire families lived and traveled together on the canals. The boats above display the distinctive bullseye emblem of the Lehigh Coal and Navigation Company, a white circle outlined with a fine blue line, with a red center. Company boats were painted a deep chestnut brown called Spanish Brown, with white trim on the gunwales and prow to increase their visibility.

Lock 3 at Bristol. *(above)* Looking west, with the upstream lock gates open and a loaded LC&N boat, behind the mule, about to be pulled in. When the boat was secure in the lock the mule could rest briefly. *(below)* Youngsters standing on closed mitre gates at Lock 3 in Bristol, with Bristol Basin behind them. The stone bollard closest to the lock shows grooves where ropes were wrapped around it. The narrow width of the original Delaware Canal locks placed limitations on the size of boats that could pass through. Most of them, like this one, were not enlarged.

*(above)* Boatmen relate that when they saw the Grundy clock, they knew they were almost at the end of their trip down the canal. Grundy Mills, a textile mill founded in the late nineteenth century, was the largest employer in Bristol; the Grundy family itself became Bristol's major public benefactor. *(below)* Loaded LC&N boat 2103 entering Lock 4, between Bristol and Tullytown.

The Delaware Canal has always been renowned for its picturesque beauty, attracting many pleasure boaters and Sunday picnickers. From the log of the Molly-Polly-Chunker, seen above near Tullytown: "Just outside Bristol we struck banks covered with luxuriant grasses and bushes, and with beautiful trees on the other side of the tow-path. We will not be apt to forget the reflections in one long calm reach of the river, with a curve at the end."

The speed limit on the canals was four miles per hour. Any faster, and the banks were damaged from the wake of the boats. A loaded boat was slower than a light boat, probably not much more than two miles per hour. Although mules did not generally need to be driven, it was common to see a boy leading a mule or a team. Haynes Road bridge can be seen in the background above. On the left is E.H. Shipman's boat with mule driver David Laury. The muzzle prevented mules from browsing on foliage along the towpath. *(below)* John Cosman, "Chief Engineer and Superintendent of Motive Power" of the Molly-Polly-Chunker, with Molly and Polly.

Loaded LC&N boats *(above)* near and *(below)* in Tullytown.

(*above*) The day was long during the canal season, 4 a.m. to 10 p.m. The mules were fed every four hours or so, and rested only while the boat was going through a lock. Captain Grant Emery's team, photographed in 1910, is unusual in that it shows a lead mule followed by a horse and a second mule. (*below*) Boats in the canal at Morrisville, opposite Trenton. Two boats are traveling together here, the front one barely visible, possibly both operated by the same family. Children younger than the boy seen here were often tied so they would not fall into the water.

(above) Partially unloaded boat at Leedom's coal yard. The two sections have been unhinged, and coal is being scooped out. The bow of the front section is being unloaded first, and the prow is riding high at the coal dock. Benjamin Klotz is standing on the deck.

(left) Louis Leedom's coal yard at Yardley.

(below) LC&N boats waiting to be unloaded at Tattersall's coal storage yard in Morrisville in 1930.

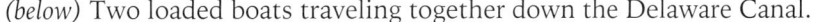

*(above)* Lock 5, at Yardley, and the locktender's house.

*(below)* Two loaded boats traveling together down the Delaware Canal.

*(above)* Near Yardley. Camelback bridges of the style seen here were the most common type of bridge along the canal. *(below)* Louis Tiffany's photo of the locktender's family, probably at Lock 6, watching the Molly-Polly-Chunker pass through the lock.

Lock 7, north of Yardley, seen from below *(right)* and *(above)* as it appeared to boatmen traveling south. Above Yardley there was a nine-mile level to the next lock. Boatmen liked long levels because they could make good time there. *(bottom)* Abandoned lime kilns at Taylorsville in 1919. Large amounts of lime and limestone were carried on the canal in boats called stone boats.

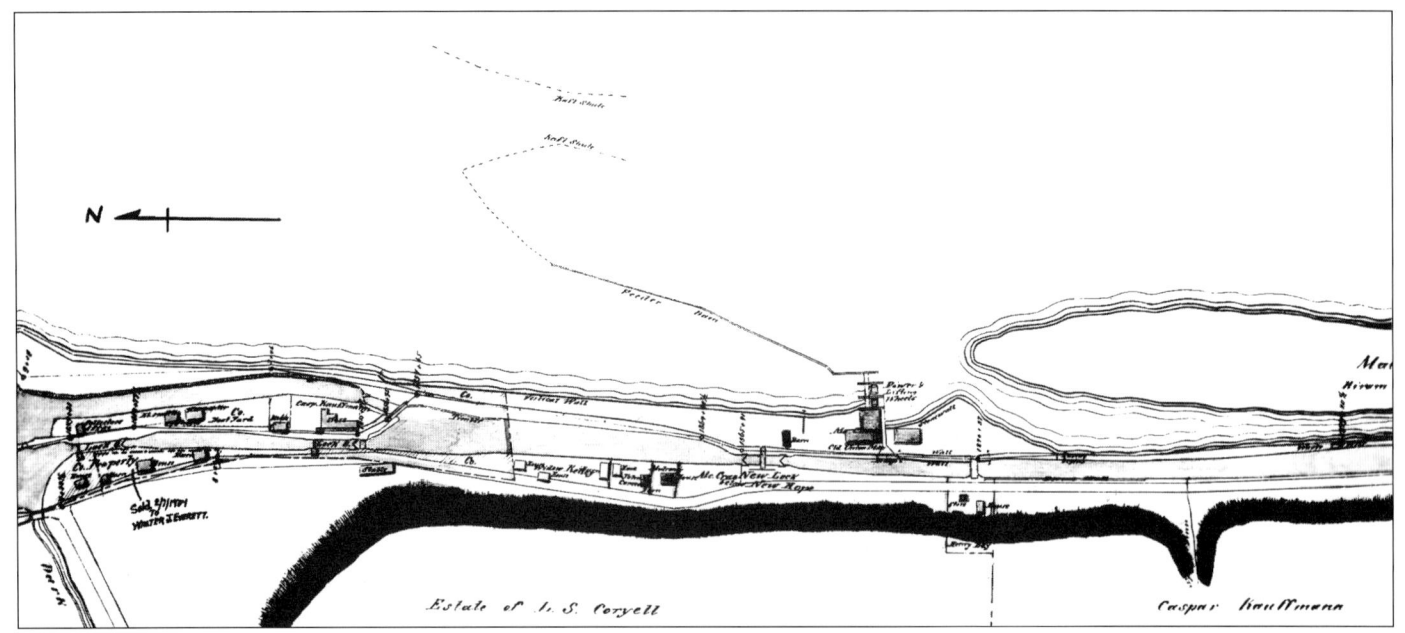

(*above*) 1868 map of the southern end of New Hope, showing the location of the lifting wheels used to pump river water into the canal. A feeder dam directs river water to the lifting wheels. In the center of the river, marked by dotted lines, is a "Raft Shute" for log rafts. (*below*) In order to increase the supply of water to the Delaware Canal, water wheels were constructed in 1832 at Union Mills, just south of New Hope, to lift water into the canal. Rebuilt in 1880, they could pump 3,500 cubic feet per minute into the canal. They ceased operation in 1923 and were largely destroyed in the flood of 1936. Today the lower end of the canal receives an inadequate supply of water from the Delaware River through the former outlet lock at New Hope.

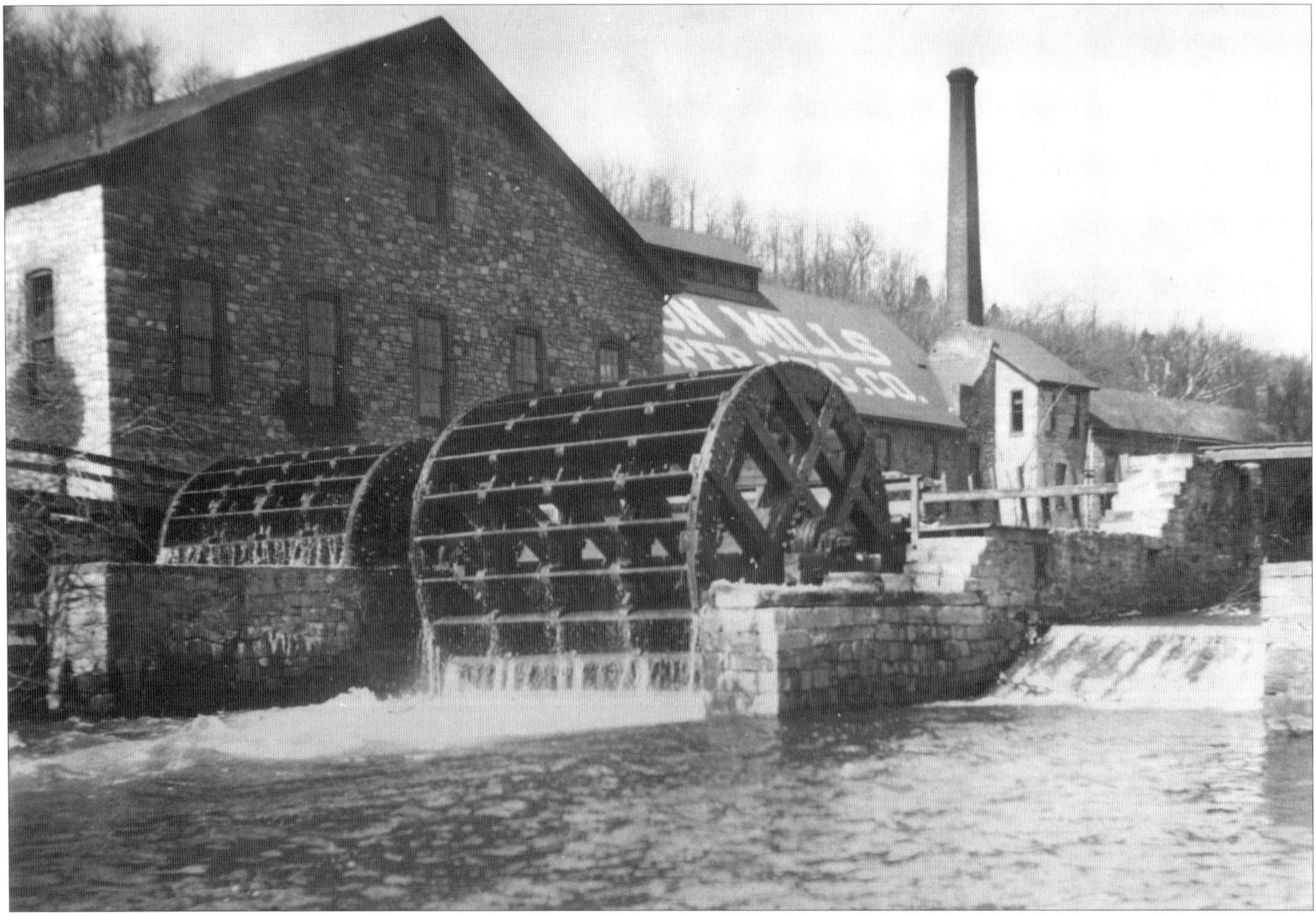

A boat entering the Delaware River from the outlet lock of the Delaware and Raritan Canal at Lambertville, New Jersey, in 1910. A cable ferry allowed boats to cross the Delaware River between Lambertville and New Hope.

## THE DELAWARE AND RARITAN CANAL

The Delaware and Raritan Canal was completed in 1834. The main canal, 43 miles in length, crossed New Jersey from Bordentown, just south of Trenton, to New Brunswick. At Trenton it was joined by a smaller branch canal, or navigable "feeder" that started at Bull's Island across from Lumberville, 22 miles north of Trenton. Boats on the Lehigh-Delaware system could transfer to this feeder after the outlet lock at New Hope was constructed in 1847. This feeder gave coal-carrying boats from Mauch Chunk a direct route to the New York City area.

The increase in traffic after the outlet lock at New Hope was completed and the cable ferry to Lambertville went into operation made it necessary to increase the length of the locks on the Delaware and Raritan to 220 feet.

The main canal brought about the industrialization of Trenton and was used by steam-powered vessels; the feeder remained a towpath canal.

The Delaware and Raritan Canal carried a total of 2,857,232 tons of cargo in 1866, its peak year. In 1871 it was leased to the Pennsylvania Railroad. During World War I it became an avenue of transportation for small vessels and military supplies that was free from the threat of German submarines.

Traffic on the canal declined rapidly after the war, and in 1923 the last boat passed over from the Delaware Canal at New Hope. In 1933 the Delaware and Raritan Canal was closed permanently to commercial navigation, and a year later it was taken over by the state of New Jersey, which uses it as a profitable water source.

*(above)* Loaded LC&N boat coming out of Lock 8 into the basin at New Hope, traveling downstream. The outlet lock to the Delaware River is on the right. *(below)* The outlet channel at New Hope, with Lambertville across the river. The cable ferry to the Delaware and Raritan Canal provided boats from the Delaware Canal with a shortcut to New York City.

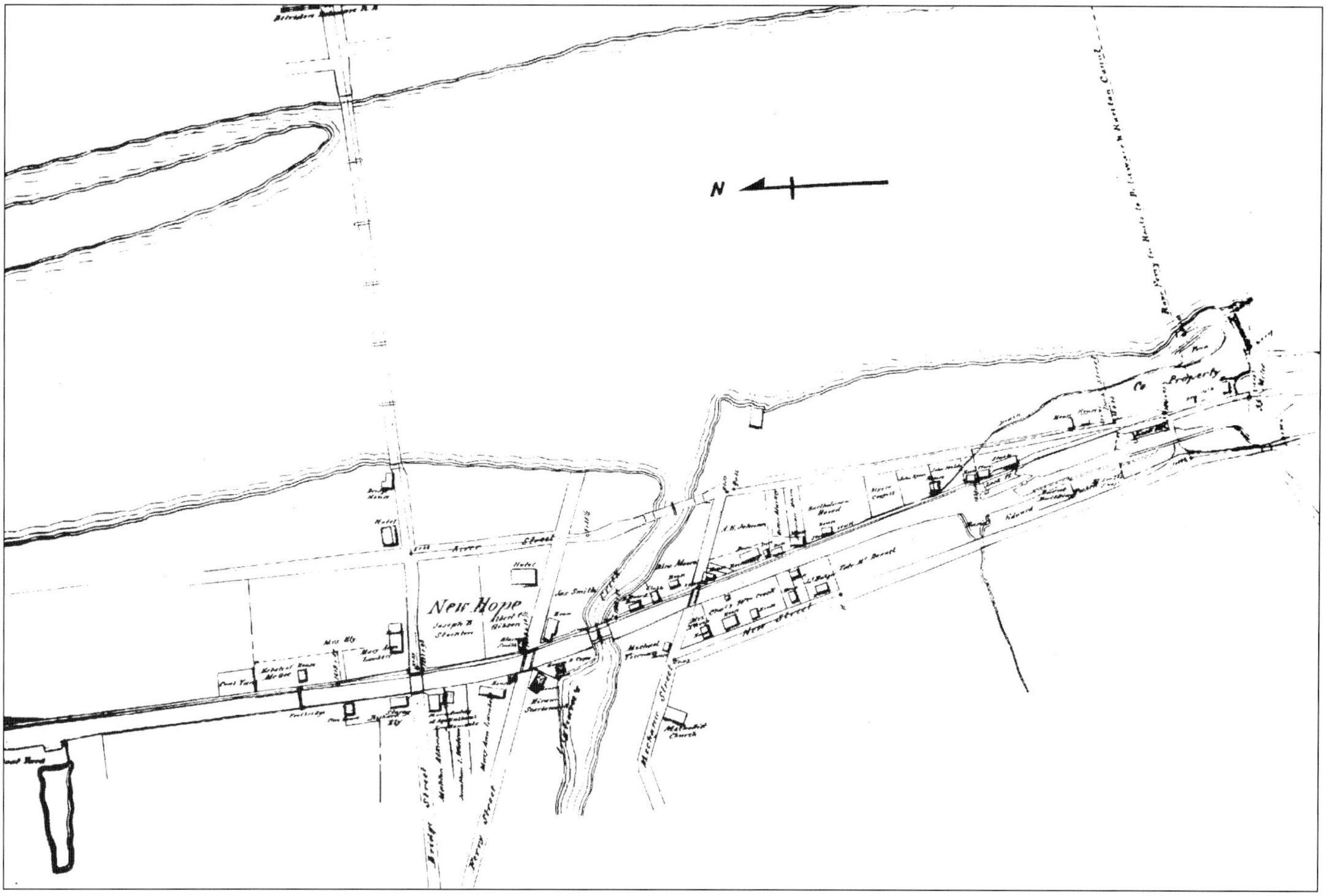

*(above)* 1868 map of New Hope, showing the outlet lock and cable ferry, also the covered bridge to New Jersey.
*(below)* Stern view of light boat No. 278 in the basin at New Hope. The rudder could be raised and lowered to improve steering. Since a light boat displaced relatively little water, the rudder would be lowered so the boat could be steered better.

(*above*) Looking south from the outlet lock toward the Union Mills Paper Company. Because of abundant water power, New Hope was the most developed industrial town in Bucks County in the late nineteenth century, with mills and businesses lining the canal. The log of the Molly-Polly-Chunker makes note of the unpleasant odors in the town. (*below*) River House, an inn and general store that catered primarily to boatmen and their families, was at Lock 8. It is now a restaurant.

*(above left)* Toll collector Samuel Sheetz in New Hope, at what is now South Main Street. Tolls were collected here from boats entering from the Delaware and Raritan Canal. *(above right)* Postcard view of a light boat tied to the side as water flows into Lock 8, a double lock. Lock 9 is in the distance. Living quarters for the crew or the boatman's family were in the small cabin at the back of the rear section. *(below)* Entering Lock 9.

New Hope had a greater change in elevation (therefore more locks) in a short distance than anywhere else along the Dela-ware Canal. In 1852, so boats would no longer be delayed too long waiting to lock through, LC&N replaced single locks 8, 9, 10 and 11 with double locks. This allowed two boats to pass through at the same time. More efficient drop gates at the up-stream ends further decreased locking times. *(above)* LC&N boats passing in the basin between locks 9 and 10. *(below)* The Lehigh Coal and Navigation Company repair yard at New Hope, between locks 9 and 10, on August 12, 1931.

*(above)* Below Lock 10, which today is under Route 32. *(below)* Two light boats, LC&N boats 278 and 150, in Lock 11. They are heading north in 1931. Double locks were 22 feet wide, just wide enough to fit two 10½-foot-wide canal boats.

(*above*) Approaching Lock 11 from the north.

(*right*) Lock 11, showing construction details on the lock wall, which has stone walls covered with heavy wooden planking. The frame house was torn down in about 1930.

(*above*) Boats 278 and 150 pulling out of Lock 11. (*below*) Postcard view of an empty "chunker" heading north through New Hope. Boatmen called all coal-carrying boats originating in Mauch Chunk (Jim Thorpe) chunkers. This one was owned by W.A. Leisenring of Mauch Chunk.

*(above)* Picnickers on an excursion near New Hope, on a Lehigh Coal and Navigation Company work boat. Canal boats never ran on Sundays, but others used the waterway provided they did not need to go through a lock. *(below)* Privately owned light boat heading north through Bucks County.

Canals had to be maintained constantly, even when there were no damaging floods. Banks and towpaths had to be kept in good repair, and sediment deposited by streams and storm-water had to be removed to keep the depth at six feet in the center. (*left*) Muddigger or shovel dredge above Rabbit Run Bridge, north of New Hope. (*below*) Work boats, including a shovel dredge (*bottom*) at work below Lumberville on August 15, 1931.

Quarries south of Lumberville, photographed in 1907 (*above*), shipped hard sandstone to Philadelphia by canal on stone boats. The sandstone was used for buildings and for paving streets. (*right*) The Tinsman family checking a raft of logs at Lumberville on a Sunday afternoon before 1903. Photo taken from the towpath. Log rafts were navigated down the Delaware River from the area above Port Jervis since the 1760s. (*below*) Part of the covered bridge at Lumberville was knocked out during the flood of 1903, and replaced with an iron truss section.

Lock 12, Lumberville. The Lumberville section was very close to the Delaware River and was subjected to serious damage when there were major floods in the river.

## FRESHETS

Canals were by nature vulnerable to floodwaters. Freshets (the nineteenth-century term for floods) could occur at any time of year. Winter freshets were the most destructive, because they were often accompanied by chunks of ice that would rip through canals and locks. Navigation would stop for weeks or months until repairs were made. The flood of January 8, 1841, caused such extensive damage to the banks of the Delaware Canal that it did not reopen until August. The most destructive flood came in early June of 1862, when booms corralling lumber in the Upper Grand Section of the Lehigh Canal broke, releasing several hundred thousand logs. An estimated one to two hundred people died, dams and bridges were destroyed, canals and locks severely damaged, and perhaps two hundred canal boats lost. Navigation was able to re-

sume by early October on the Delaware Canal, leaving very little time for winter coal supplies to be stored. LC&N decided at that time to construct a railroad to replace navigation on the Upper Section.

Troublesome freshets occurred, locally and valley-wide, with unpredictable frequency. No coal was shipped at all during 1902 because of floods on December 14, 1901, and March 2, 1902. While repairs were being completed the Delaware River flooded again, on October 10, 1903, causing extensive damage to the Delaware Canal. Railroads were transporting most of LC&N's coal by this time, but there were many customers along the Delaware Canal whose coal yards could be served only by canal. The company kept repairing flood damage until the decision to close the canal in 1931.

*(above)* Private boats left in the road at Delaware Quarries, near Lumberville, by the flood of 1903. This flood hit the Delaware Canal very hard, but was not serious on the Lehigh Navigation, whereas the 1902 flood had been disastrous on the Lehigh and less damaging on the Delaware. *(below)* Company boats tossed about in the canal by the 1903 flood.

*(above)* The lock at Lumberville following the flood of 1936.

*(left)* A break in the towpath, 86 feet long and 20 feet deep, at the two-mile level one thousand feet above the lock at Lumberville in August 1912 was repaired immediately. The crib with two rows of sheeting is considerably stronger than the dirt bank it replaced.

(*above*) Private light boat heading upstream coming out of Lock 13. (*below*) Two private boats docked at the coal yard at Point Pleasant. Private boats were higher than LC&N boats, and had peaked hatch covers. The long boom is a derrick used to unload coal; coal was shoveled manually into a 200-pound-capacity bucket on the end of the derrick, which was then swung over and emptied at the coal yard. Coal yard owners often paid boatmen $10 to $20 to shovel the coal, an unpopular job generally disdained, at least in later years, by locals.

(above) When boats passed, the loaded boat would slow down and move to the berm side of the canal, opposite from the towpath. As it did this, its tow rope, heavy because it was saturated with water and grit, sank to the bottom of the canal so the light boat could ride over it. Robbie Best of Walnutport and Howard Reed of Freemansburg are the captains of these two boats. Martha Best, in the loaded boat with wash on the line, is steering her boat to the berm side. (below) Lock 13, the lower lock at Point Pleasant, in a postcard view that shows the overflow on the left. (right) Lock 14, the upper lock, in a 1930 Francis B. Palmer photograph.

Camp Brumbaugh,    LOCKS ALONG THE DELAWARE CANAL.    Point Pleasant, Pa.

*(above)* Postcards were made of numerous picturesque spots along the Delaware Canal. This one was possibly taken on a Sunday, as the canal was closed to commercial traffic that one day of the week. Boats can be seen lined up along the banks. *(below left)* The Tohickon Creek aqueduct, a popular local beauty spot, was seriously damaged in 1934 *(right)*. The temporary repair, a wooden flume, was destroyed in the flood of 1936 . The aqueduct was rebuilt in 1948.

*(above)* A group of picnickers on a work scow heading toward Lock 15–16, returning from a trip to New Hope. *(below left)* Inspecting the Delaware Canal on the two-mile level above Point Pleasant in 1901 are, left to right, J. M. Line, John W. G. Young, and I. M. Church, the canal superintendent from 1898 to 1930. Church understood boatmen and was loved and respected by them. *(right)* Jacob H. Henry, locktender at Lock 15–16, showing Sea Scouts how to operate the lock gate. Treasure Island, a camp owned by the Boy Scouts of Philadelphia, is nearby.

(above left) Flora Henry first came to Lock 15–16, the Smithtown Lock, as a child of five in 1915, when her father, Jacob Henry, was appointed locktender there. She assisted him until his death in 1931, then worked there alone until the canal closed soon after. The double lock was 12 feet deep, the second deepest on the canal. The photo of Flora Henry was taken by C. P. "Bill" Yoder, author and canal enthusiast who wrote *Delaware Canal Journal*, a history of the canal and its people. (above right) Lock 15–16, August 6, 1931. (below) A work scow and a carpenter boat entering Lock 15–16.

Jacob Oberacker's tap room, now the Golden Pheasant Inn. One entrance faced the canal, another faced River Road. The black boatman on the towpath is Jimmy Brown, one of the last men to run boats on the canal.

## PROVISIONS

Many stores along the Delaware and Lehigh canals supplied boatmen with what they needed. The boats had only a wet well and no proper refrigeration to keep food fresh, so daily supplies for the boatmen, their families and their mules could be bought en route. Some supplies were stored on the boat; smoked meats such as ham and bacon could be kept safely insulated in the feed box, but perishable foods were bought or bartered daily. Along more isolated stretches of the Lehigh Navigation, such as Hope's Lock and Chain Dam, locktenders or their wives and childen would operate small stores at the lockhouses, but since River Road parallels most of the length of the Delaware Canal, general stores and taverns developed to serve canal traffic, local residents, and travelers on what was then the main road between Easton, Lower Bucks County, and Philadelphia.

A number of the early inns and taverns are still in operation as restaurants. Only one general store is still in business.

*(above)* Family heading south on the Delaware Canal around 1900. Their boat is a stiff boat, a type that was not common and fell out of use by 1920. They were harder to handle than section boats as they could not be broken apart for loading, unloading, and turning around. Two young children are following the mules, which are feeding. *(below)* The Tinicum Creek aqueduct on August 8, 1931. On the left is the covered bridge that carried River Road over the creek.

*(above)* Looking north from the camelback bridge at Erwinna. *(below)* A company boat and two private boats moored in the center of the canal at Erwinna, awaiting the removal of water for the winter. A number of boatmen had homes nearby, where they would spend the months December through March, or however long the canal was closed because of ice. Only a small amount of water was left in the canal during the winter, otherwise boats could be damaged by ice. Maintenance on the banks was performed during this time, and muskrat holes were plugged. The bounty offered by LC&N on muskrats helped keep them under control, or they would have drained the canal.

(above) Looking south toward the cam-elback bridge at Erwinna. (right, below and opposite page) The unidentified trav-elers who had started in Bristol several days earlier. The mule is draped with cloth to keep off biting insects.

Uhlertown Boat Yard

The village of Uhlertown became a thriving canal community in the 1850s, when Peter Uhler established a number of businesses there including a grist mill, lime kilns, and a coal yard. He built canal boats and ran his own line on the canals. *(above)* The boat yard, on the right bank downstream from the covered bridge. *(below)* Loaded boat No. 124 passing by the Uhlertown boat yard.

*(above)* The Uhlertown Hotel is in the center of the picture; the dark building is the locktender's house for Lock 18. *(right)* Lock 18, and the Sigafoos store and home. *(below)* The covered bridge over the canal at Uhlertown has been in continuous use since 1832 and is the last remaining span of this type on the Delaware and Lehigh canals. On its left is the mill as it appeared in 1917.

(above) Locking down at Lock 19 at Lodi. Single locks, 11 feet wide and 95 feet long, were the original, unaltered locks built by the state. They retained miter gates at both the upstream and downstream ends. Boats were six inches narrower than the locks, and almost 90 feet long. Near the front of the boat a nighthawker (lantern) is hanging on the dasher, which is decorated with the words "Star of Bethlehem." To its left, a bilge pump is lying on the deck. The tow rope with a knot on its end was slipped through the notched towing post on the left and could quickly and easily be released when necessary. (right) Loaded boat 2152 traveling downstream near Lodi, with light boats tied up on the side. (below right) The tainter gate at the waste weir at Grey's Creek, just north of Lock 19, released excess water from the canal to prevent overflows. The Delaware Canal State Park headquarters is at Lock 19.

*(above)* Louis Tiffany photograph of the boatyard at Upper Black Eddy, also known as Bridgeton. *(below)* The boatmen's entrance to the general store at Upper Black Eddy, the only one along the canal still doing business. Known during most of the canal era as the Singley Store, it was earlier run by Mike McIntee. It was a social and commercial center for boatmen. Many captains tied up for the winter nearby, as more boatmen had homes in the Bridgeton area than anyplace else.

(above) While Molly and Polly are taking a break, photographer Louis C. Tiffany attracted the attention of these children in Bridgeton. (left) Two cameras were carried on the Molly-Polly-Chunker, enabling Tiffany and his friend Walter C. Tuckerman to capture a curious matron in Bridgeton. The photographers had to carry not only their bulky cameras on the trip, but also a makeshift darkroom and their chemicals. (below) Market gardening along the canal, one half mile north of Lock 19. August 15, 1931.

*(above)* A group of pleasure seekers on a scow above Lock 20 at Narrows.
*(below)* A stiff boat carrying excursionists through the Palisades, approaching Lock 20.

(left) On passing through the Palisades, the high cliffs at the Narrows, the log of the Molly-Polly-Chunker noted: "The scenery this day was by odds the most beautiful we had seen, so far." (below) House boat entering Lock 20 at Narrows. (bottom) Loaded boat approaching Lock 20 from the north. After the canals were closed, this section of the Delaware Canal was threatened with being filled in to widen narrow River Road.

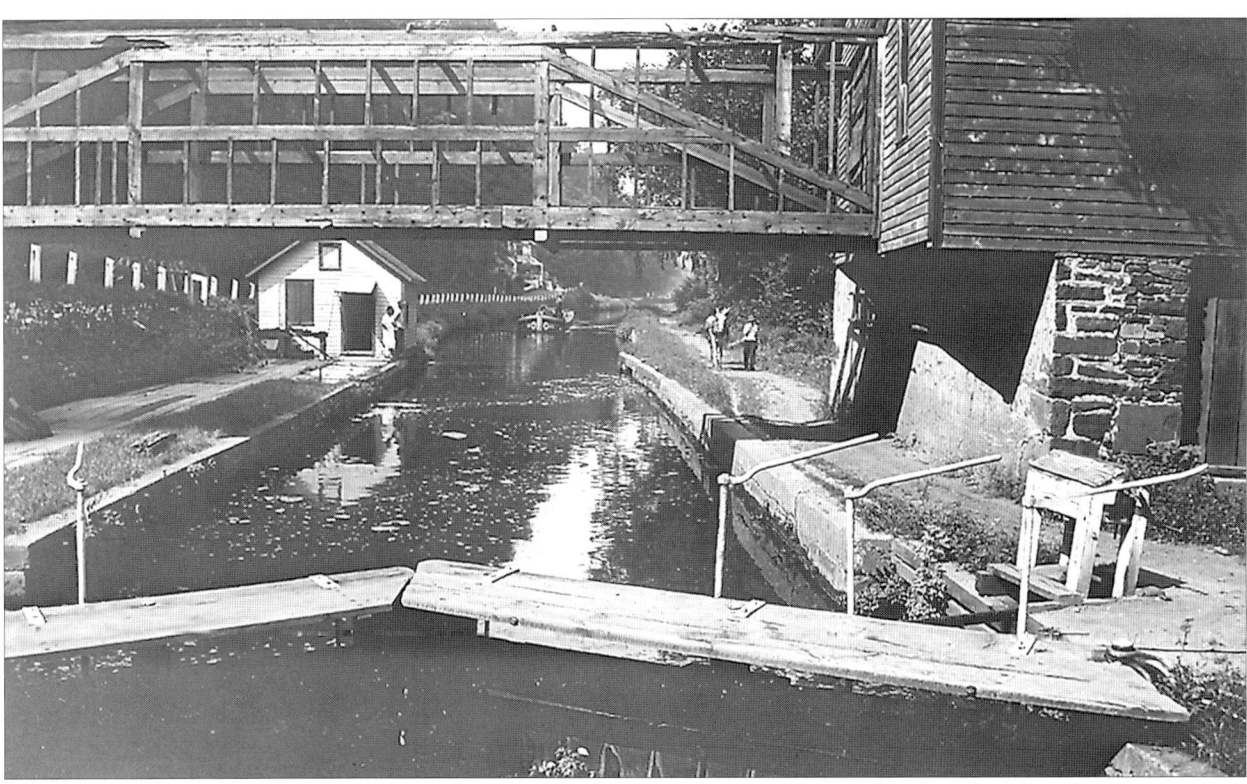

(above) Louis Tiffany's photograph of a family living alongside the canal, probably in the Kintnersville area. (below) Loaded boat approaching the lock below Kintnersville.

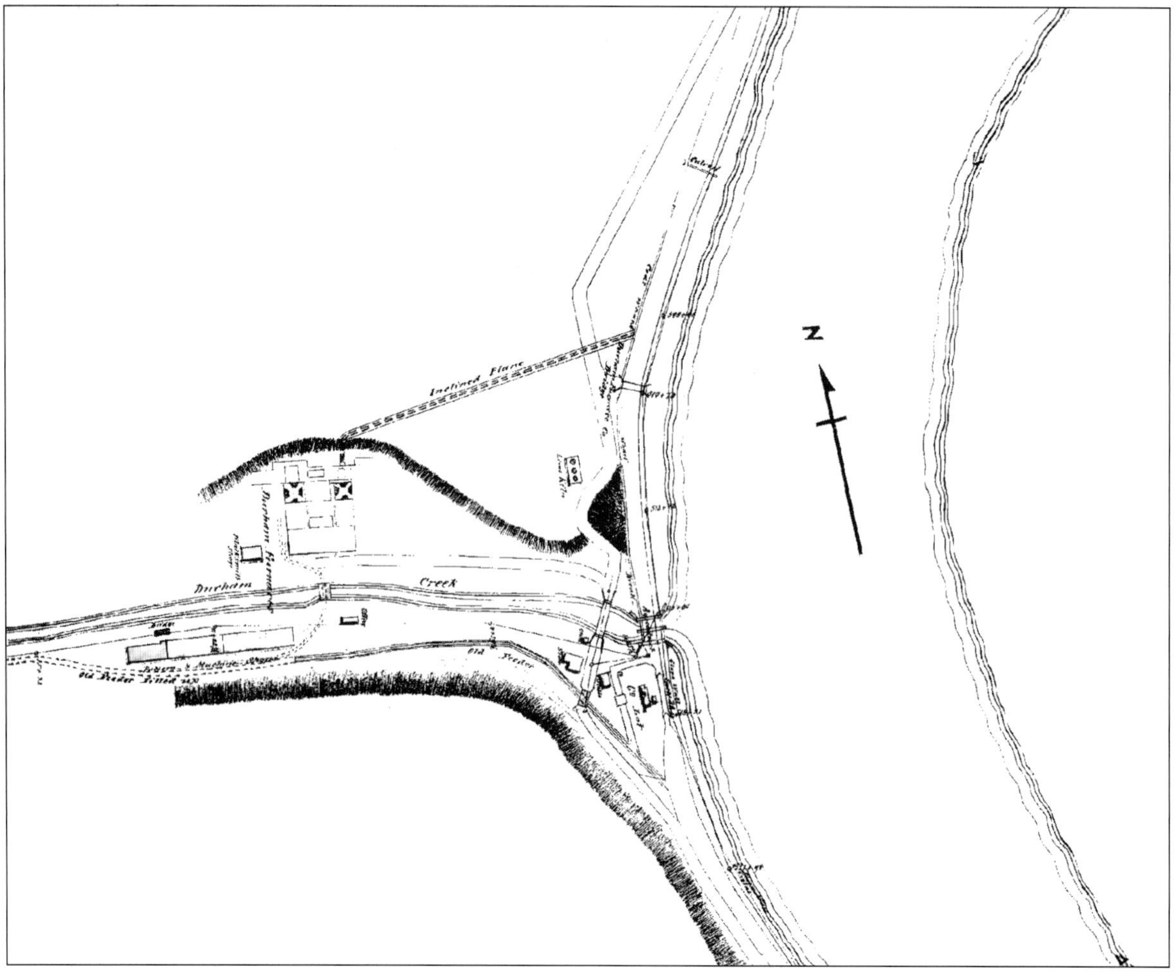

Two anthracite-fueled blast furnaces were built near the canal on the Durham Creek in 1848–49. Anthracite was brought in by boat and taken to the furnace on an inclined plane from the wharf to the stockyard above the furnace; pig iron was shipped out on the canal. The Durham Furnace, for many years operated by the firm of Cooper and Hewitt, was the only furnace along the Delaware Canal. It shut down in 1908. *(below)* When members of the American Institute of Mining Engineers visited Durham in 1886, part of their tour was by canal boat.

*(above)* The aqueduct above Lock 20 at Durham. *(below)* Postcard view of the Delaware Canal at Riegelsville.

*(above)* LC&N boat No. 286 heading upstream through the village of Riegelsville. *(below)* In the background along this stretch in Riegelsville is one of the eight stop gates on the canal. They were installed to minimize flood damage by diverting floodwaters into the nearby Delaware River.

(above) At Raubsville a hydroelectric plant used canal water until 1954 to produce electricity for the Philadelphia and Easton Transit Company and the Raubsville Paper Company. When the canal was given to the state in 1931, the Lehigh Coal and Navigation Company retained control of the section between Easton and Raubsville because of the substantial revenues it received from the sale of water. (below left) Boat No. 281, seen from the bridge at Raubsville. (below right) The canal, the road, and the Philadelphia-to-Easton trolley below double Lock 22–23. The tailrace from the hydroelectric plant is entering the canal in the foreground.

(*above*) Fred Walck is standing on the deck of the scraper boat that maintained the section of canal from Easton to the Raubsville lock; Clayton Dotter is at the window. The boat is under the railroad bridges at the Guard Lock at Easton. The Raubsville lock, No. 22–23, was known as Groundhog Lock. It was a combined double lock with a lift of 17.3 feet, the highest on the Delaware Canal. (*below*) View from the bluff below Easton, looking toward Easton. with Phillipsburg on the right, where the coal-loading docks at Port Delaware on the Morris Canal can be seen.

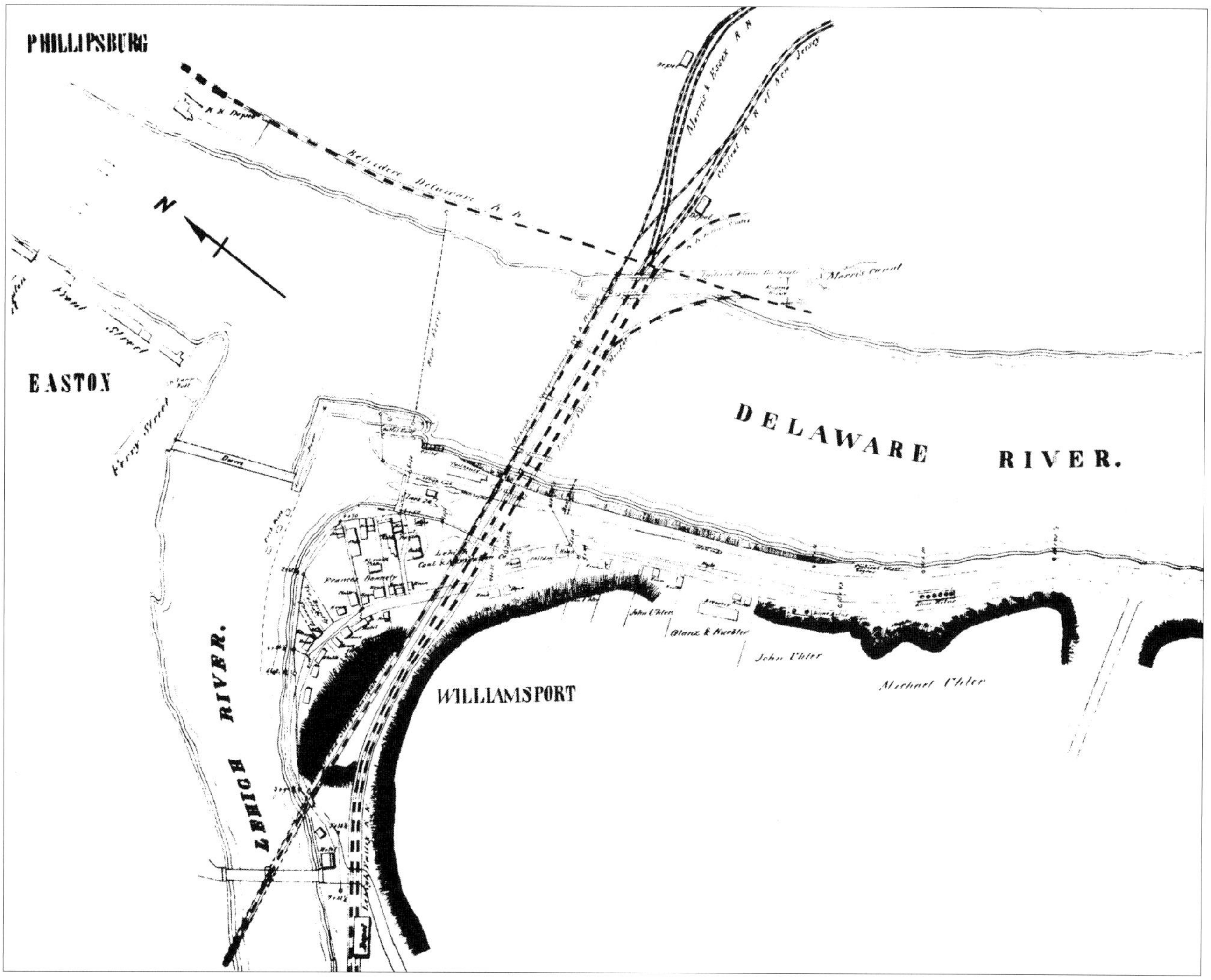

(above) The end of the Delaware Canal, the beginning of the Lehigh Navigation, and the entrance to the Morris Canal can all be seen in this 1868 map of the Forks of the Delaware. The land on the right bank of the Lehigh River at its confluence with the Delaware was intensively developed with canal-related businesses. Its official name was Williamsport, but it was known locally and by boatmen as Snufftown.

(right) Local children watch the Molly-Polly-Chunker pass through the guard lock at Easton, and out of the Delaware Canal.

Phillipsburg is on the right and Easton in the background, with Williamsport (Snufftown) and South Easton on the far left. The Delaware Canal ended under the railroad bridge in Williamsport on the opposite side of the river from this vantage point. From there, boats could lock into the Lehigh Canal or cross the river by a cable ferry to the Morris Canal. The entrance to the Morris Canal is under the railroad bridges in the center of this pre-1891 photograph. The Lehigh Valley Railroad and Central Railroad of New Jersey bridges crossed the Delaware at the same point. The coal-filled cars are on the Lehigh Valley tracks.

## THE MORRIS CANAL

The Morris Canal, completed between Phillipsburg and Newark in 1831, provided boats on the Lehigh and Delaware canals with their first shortcut to the harbor in New York. Entirely different from the Lehigh and Delaware waterways, it used a system of unique water-powered inclined planes to cross the New Jersey hills. In addition to the 23 planes, there were 32 lift locks along the 102-mile canal. Built originally with a ten-ton limit, the canal was upgraded by 1845 to accommodate section boats carrying 44 tons. An 1860 improvement program increased the capacity to 70 tons.

The primary freight on the canal was anthracite coal shipped eastward from Pennsylvania. However, there was significant westward traffic in iron ore for the iron furnaces in the Lehigh Valley, which typically used one-quarter New Jersey ore mixed with local limonite ores in each blast.

The Morris Canal's peak year was in 1866, when it carried 899,220 tons. Increasing competition from railroads resulted in a rapid decline in volume of traffic. In 1871 it was leased by the Lehigh Valley Railroad, which constructed the coal-loading chutes at Port Delaware. Coal was brought there by train and transferred to canal boats for further shipment via canal. But traffic continued to decline; several attempts, the first in 1903, were made by the state legislature to abandon the canal until in 1922 most of it was taken over by the state. Between 1924 and 1927 the canal was abandoned and largely destroyed. Its feeder lakes and the Hudson River basin became public property.

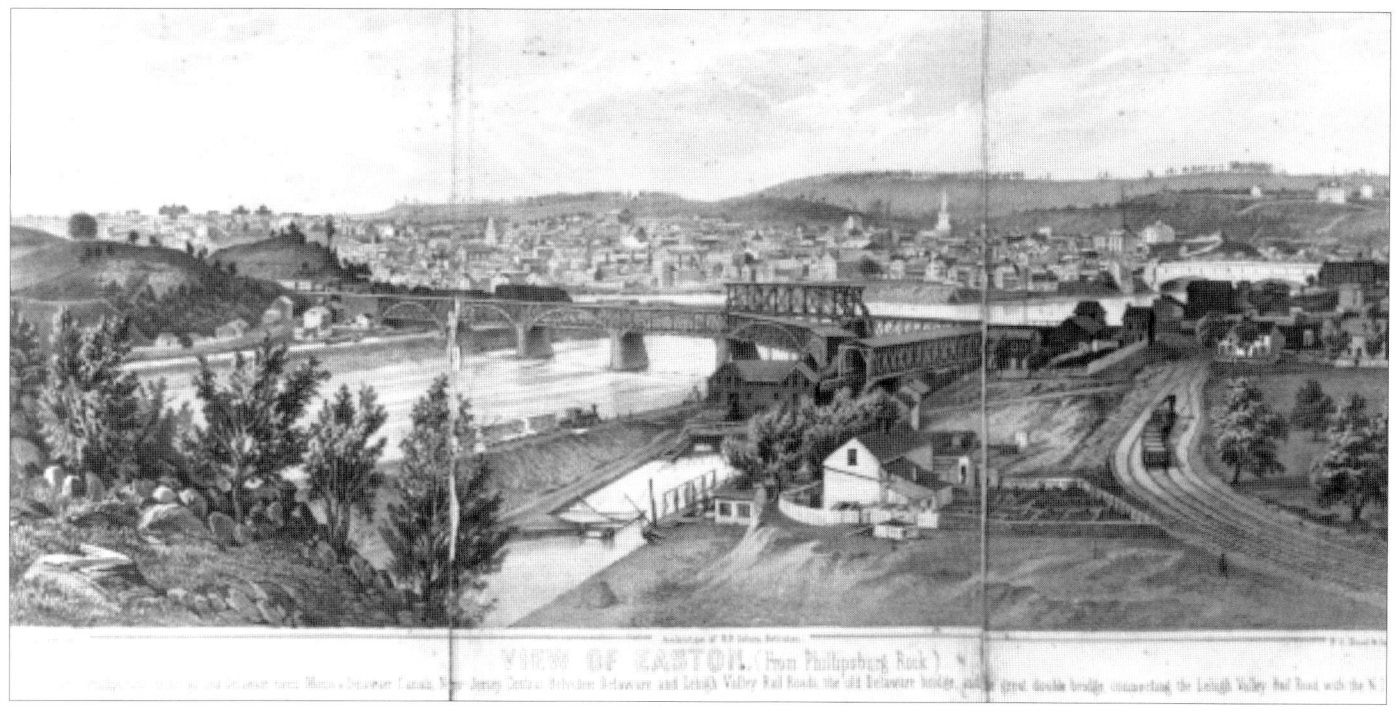

VIEW OF EASTON. (from Phillipsburg Rock)

*(above)* Pre-1871 lithograph of the Easton-Phillipsburg area, before the Lehigh Valley Railroad developed coal-loading docks at Port Delaware on the Morris Canal. *(below)* View, circa 1904, from Phillipsburg over the Delaware River south of Easton. The upper end of the Morris Canal boat basin, known as Port Delaware, and the Lehigh Valley Railroad coal chutes can be seen. Across the river, on the bank of the Delaware Canal in Pennsylvania, is a long row of large limekilns.

(*above*) Empty section boats at Port Delaware in 1891, waiting to be filled from the coal chutes on the left. The chutes were built after the Lehigh Valley Railroad took over the Morris Canal in 1871. Coal was transported by train to this point, then transferred to canal boats to be shipped to the New York City area. The cable ferry across the Delaware River was no longer used after the mid-1880s. (*right*) Morris Canal boat #486 west of Lock #8 West on August 8, 1897. Morris Canal boats measured 93 feet long, 6 feet high and 10½ feet wide.

# THE LEHIGH NAVIGATION

At Easton, the canal changes. The first large concentration of industry, railroads, and people is here, at the Forks of the Delaware. Along the waterfront were warehouses and factories served by canal boats. An early industrial park and large iron furnaces were developed in the South Easton–Glendon area, using the canal for both waterpower and transportation. Further up the canal, there were many more small and large industries.

The Lehigh Canal differed greatly from the Delaware Canal in several respects. It passed through an urban and industrial landscape, while the Delaware passed through a landscape that was largely pastoral; its locks, twice as wide as most of those on the Delaware, could accommodate two boats at one time; and it combined slackwater and canal sections, whereas the Delaware was a true canal, separated from the river for its entire length.

Boatmen preferred the Delaware Canal to the Lehigh Navigation because they could make better time, and time was money. In 60 miles, the Delaware had 24 locks to overcome the drop of 180 feet. The Lehigh, on the other hand, dropped 353 feet in only 46 miles, and needed 44 lift locks, five guard locks, and three guard

lifts. Furthermore, the towpath changed from one side of the river to the other, a time-consuming maneuver that boat captains did not like. So long as the canal was level, a boatman made good time, but when he had to go through locks and cross the river, the pace slowed.

The Lehigh Coal and Navigation Company retained virtual ownership of the Lehigh River until as late as 1966–1967. Then, through the efforts of Allentown's state representative, Sam Frank, control of the river was returned to the state to be administered by the Department of Environmental Resources. The company gave or sold the canal and right of way to local governments, sportsmen's clubs, and private landowners.

By the time the company relinquished control, many sections were irretrievably lost and most of the dams were destroyed. Since then, more dams have gone, releasing tons of accumulated coal silt that is slowly traveling downstream. Today, the river has canal dams only at Easton, Chain Dam, and Allentown, and portions of the canal have been filled in, paved over or washed away. Of the stretches that have been preserved and are now open for public enjoyment, the only section that could still be used by canal boats is the two and a half miles from Chain Dam to the Easton dam.

*(top)* Captain John Best and his wife Martha "Hautie" Best moving their boat from the Delaware Canal into the Lehigh Navigation. *(below left)* Lehigh Coal and Navigation Company boat 213 traveling light up the slackwater past Mount Ida in Williamsport (South Easton) in 1910. Miss Billie Burke in "Mrs. Dot" on Broadway is featured on one of the advertising posters above the towpath. *(below right)* Lyman H. Bagg, alias the Wheelman, a friend of the company of the Molly-Polly-Chunker, watching activity on the canal from the north bank, across from Mount Ida.

(*above*) Forks of the Delaware, with much of Easton on the right. The Lehigh Canal is more appropriately called the Lehigh Navigation, as it was a combination of canal and slackwater pools behind dams. (*below*) The Easton dam and waterfront. This dam raised the level of the Lehigh River at its confluence with the Delaware, thus providing water for the Delaware Canal.

*(left)* Looking downstream toward Guard Lock 24 of the Delaware Canal. The Third Street bridge and the Central Railroad of New Jersey bridge cross the Lehigh at the same point. The towpath is on the right. *(below)* One of two pedestrian suspension bridges over the Lehigh River and Canal, this one at Abbott Street was built in 1886 so workers could cross to the Lehigh Valley Railroad shops. It was blown down in a storm in 1951. The LVRR roundhouse is prominent in the center. Circa 1900. *(bottom)* The Abbott Street suspension bridge in about 1890, viewed from the north, showing a variety of industrial enterprises along the waterfront.

(above) Dock Street in Easton, where numerous warehouses were built along the slackwater navigation. The Central Railroad of New Jersey is elevated above the street. (below) The Ingersoll Rand Company, seen from South Easton with the canal in the foreground, about 1900. The Ingersoll Sargent Drill Company's plant was in operation by 1895. Ten years later it became part of the Ingersoll Rand Company.

Two postcard views of the Lehigh River at Easton. *(above)* Looking upstream in 1908. The original outlet lock where descending boats left the canal to enter the slackwater pool behind the dam at the mouth of the river is in the left center. *(below)* Looking from Easton over the Lehigh River and canal to South Easton in the 1890s. The bridge is that of the Easton and Northern Railroad. Part of the Abbott Street industrial area is on the right; the original outlet lock is on the far left.

*(above)* On the left is the Abbott Street area circa 1900, showing the boat basin and original outlet lock. In the far left background is the South Easton furnace of the Glendon Iron Company. In 1904 the basin was removed and the outlet lock moved up the canal to make room for expansion of the trackage of the Lehigh Valley Railroad. *(below)* The Abbott Street industrial area, circa 1885. Begun in 1832, Abbott Street was one of America's first industrial parks. It was developed by the Lehigh Coal and Navigation Company because of the abundance of water in the canal, which the company sold to power mills and factories. By 1850 over one thousand men were employed there making such products as rifle barrels, whiskey stills, cotton, iron wire, and sawn lumber. By 1900 almost all of these industries had ceased operation.

*(right)* Reconstruction of the outlet lock at South Easton in 1904, showing placement of the drop-gate wickets, which were salvaged from the original lock. The lock was rebuilt using concrete.

The Glendon Iron Company in 1890. The canal boat is an example of boats LC&N built to operate only on the Lehigh Navigation. They had a greater capacity than standard coal boats, but they were too wide and high to navigate the Delaware and Morris canals. They operated between the coal-loading docks and the iron furnaces of the Lehigh Valley, which were heavy consumers of anthracite.

## GLENDON IRON COMPANY

The Glendon Iron Company was the second company to build anthracite-fueled furnaces on the Lehigh Canal. The first of its four furnaces went into blast in 1844, just four years after anthracite was successfully used for the first time in the United States to make iron. Iron was produced using anthracite for fuel in 1840 at Catasauqua by the Lehigh Crane Iron Company, a company founded by the owners of the Lehigh Coal and Navigation Company, principally Josiah White and Erskine Hazard. By 1873 there were 47 anthracite-fueled furnaces in the Lehigh Valley, producing 389,967 tons of iron, by far the greatest amount of any region of Pennsylvania or the nation. That year, 49 percent of all iron produced in the United States came from Pennsylvania.

Until 1855, all the furnaces were dependent on the Lehigh Canal for anthracite, and were built at the canal's edge. After 1855, railroads emerged as the principal coal and ore carrier and furnaces began to be erected in places convenient for rail transportation.

The proximity of the plant of the Glendon Iron Company to the canal and the Lehigh Valley Railroad can be seen above. All the early anthracite furnaces in the Lehigh Valley used local limonite ores; in the case of the Glendon Iron Company, the primary ore was brown hematite from adjacent Williams Township, which was supplemented by magnetic ores from New Jersey. The magnetite was shipped on the Morris Canal to the Lehigh Canal.

Until 1857, pig iron from the Glendon furnaces was shipped out on the canal system to New York where it was loaded onto coastal schooners to be taken to the Glendon Rolling Mill in Boston. After 1857 the company concentrated on producing "grey" iron for use in forges. The works were shut down in 1894.

(*above*) The Lehigh River and Canal seen from Glendon. Chain Dam and Island Park, an amusement park in the middle of the river, can be seen in the background. In the center is the site of the Lucy Furnace, torn down in 1897 and now part of Hugh Moore Park. Lucy Crossing bridge was one of several covered bridges over the Lehigh Canal. (*below*) The *Josiah White*, seen here in 1979 with mules Jessie and Jennie led by Vickie Coyle, carried tourists on the canal through Hugh Moore Park from 1978 to 1992. The team of Dixie and Daisy started to pull the boat in 1991. The National Canal Museum launched a larger boat, the *Josiah White II*, on November 4, 1993; its inaugural cruise was in April, 1994. A new mule team was introduced in 2006.

(*above*) Postcard view of Chain Dam and the guard lock at the entrance to the Glendon–South Easton section of the canal. The locktender's house stands on the narrow strip of land between the river and the canal. The house is preserved by the National Canal Museum as a living history museum. (*below*) Canal workers at the Glendon guard lock above Chain Dam, photographed on August 11, 1886.

(above) Chain Dam, at the upstream end of the Glendon level. The lock-tender's house seen in the photo opposite has not been built yet. The original one burned down in the 1920s. (right) Work boat in the canal at Chain Dam in front of Keystone Furnace about 1900. The small stone building is the original locktender's house for the guard lock. (below) A company scow at Guard Lock 8, just above Chain Dam, with the new lock-tender's house visible.

(*above*) No. 5 dredge of the Lehigh Coal and Navigation Company in Section 8, the Glendon level, in 1901. Workers' housing built by the Glendon Iron Company during the 1840s is in the background. This is the section used by the National Canal Museum for mule-drawn boat rides. (*below*) Change Bridge, above Chain Dam. The bridge connected the south bank of the Lehigh River and the island, which was formerly Smith's Island and from 1894 was known as Island Park. Mules walked across Change Bridge to cross the river. On the north side of the island they completed the crossing on a causeway. Change Bridge was built in 1857, using wire rope from LC&N's wire-rope factory in Mauch Chunk, built in 1847–48 and the first such factory in the United States. An amusement park was developed on the island starting in 1894, with access by a trolley line from the north side. In 1919 an ice floe destroyed the trestle for the trolley line, and the amusement park was abandoned. Today the island is a wildlife sanctuary, part of Hugh Moore Park.

(above) A group of ice cutters at the ice house on Island Park, circa 1895. The ice was taken from the Gut, the bay at the downstream end of Island Park, where fresh springs entered the river and water froze readily because of the lack of current. Change Bridge can be seen on the far left. (below) Making repairs to Chain Dam on August 27, 1932.

(*above*) Light section boat moored, possible on a Sunday when the canal was closed. Boatmen lived on their boats, carrying supplies for at least part of the seven- to eight-day trip from Mauch Chunk to Bristol and back. Their small living quarters were at the stern of the boat. (*below*) Loaded boat moving downstream through Hope's Lock. The cast-iron bollard on the right, used to snub boats as they were entering the lock, was known by boatmen as a "nigger head."

Swimming was a popular activity in the canals, although the coal silt could be so bad that swimmers came out of the water speckled with black. Among the group above are the Gessler sisters. Clubs and camps were established along the canals. The Gahuwa Club, which became the Northampton Country Club, was in the Hopesville area, across the river from Bethlehem Steel's huge complex at Redington. *(below)* Standing near the clubhouse of the Gahuwa Club in 1911 are, left to right, an unidentified boy, Arch Johnston, Jr., Bob Stout, Tom Loeser, and John Loeser. Arch Johnston, Sr., was vice president of Bethlehem Steel under Charles Schwab, and the first mayor of the consolidated city of Bethlehem.

(above) Lock 44 at Freemansburg. The house is one of only two original stone 1829 locktender's houses left on the Lehigh Canal. The other is at Walnutport. (below) Boat 200 has just moved out of Lock 44, heading downstream. The mule driver on the towpath is Joe Reed.

(above) Geissinger's mill on the canal at Freemansburg was powered by an undershot wheel in the Lehigh River, and shipped some of its products by canal. (below) Looking toward the stern of boat 227, docked at Freemansburg, can be seen the iron-plate cookstove with its chimney chained to the boat; in the center are the coal box and shovel. On the right is the water barrel and the notched towing post. These items are on the front section of the boat. On the rear section is the feed box for oats and corn, in which boatmen would store their ham and bacon to keep them cool.

*(above)* The view from Pop Rock, Freemansburg, toward Bethlehem and the steel works. Along the railroad tracks can be seen the Schimer Foundry, manufacturers of cast-iron toys and decorative ironwork. *(below)* The Bethlehem Steel Company in about 1907, photographed from the bank of the canal by William Rau. The Bethlehem Iron Company, which became Bethlehem Steel in 1899, was built by railroad interests. Unlike the earlier, anthracite-fueled furnaces, it was never supplied by nor dependent on the canal.

(above) Lock 42 at Bethlehem. (below) The station of the Central Railroad of New Jersey in Bethlehem, and the canal. In the forefront is the Keystone House, which later became the main office of the Fritch Coal Company. The Moravian settlement was adjacent to this section of the canal.

## BETHLEHEM

Bethlehem did not develop in response to the construction of the canal as readily as did other towns in the Lehigh Valley. The town was a closed religious community, and when the canal was built there was much resentment in the Moravian community for aesthetic and practical reasons. The dams downstream cut off their supply of shad, a major source of food, and the canal was viewed as a scar on the landscape. The river bank and the mouth of the Monocacy Creek were altered, affecting their water-powered industries.

The canal was an intrusion into the way of life of the Moravians; they knew it would bring change and open up the Bethlehem market to outside competition, which they did not want. Furthermore, typhoid fever broke out while the canal was being dug in Bethlehem, which the people blamed on the freshly dug earth.

(above) The northern end of the New Street bridge in 1882, with LC&N coal cars on the Lehigh and Susquehanna Division of the Central Railroad of New Jersey. The company's bullseye emblem appeared on both its canal boats and its coal cars. (below) A later photo of the same location, showing a private boat under the bridge, approaching Lock 42.

*(above)* Looking west from the New Street bridge to Lock 42. The tracks are those of the Central RR of New Jersey. *(below)* LC&N boat No. 202 in Lock 42, heading downstream. Locks on the Lehigh Canal were 22 feet wide, twice as wide as most of the Delaware Canal locks. They could accommodate two boats simultaneously.

(*above*) Looking from the New Street Bridge toward Lock 42. This photo shows new development along the railroad that was not there in the photo on page 89. The first building on the right of the tracks belongs to the Allentown-Bethlehem Gas Company. Further down is a large sign reading "Davies-Strauss Stauffer Co. Wholesale Grocers." (*below*) View in 1889 from the New Street Bridge.

## TRANSPORTING ANTHRACITE

The Lehigh and Delaware canals were first and foremost coal carriers. Built initially to carry anthracite coal to Philadelphia, they became a conduit for coal shipped all over the eastern United States, particularly after connections with the Morris Canal and the Delaware and Raritan Canal were completed.

Anthracite was a prized fuel for as long as householders cooked and heated with coal stoves, industry ran on steam engines, and trains pulled by steam engines (often fueled with bituminous) carried the goods and people of the nation. In the mid-nineteenth century the canals were, or at least appeared to be, a solid mass of boats, almost all of them carrying coal. Countless small and large coal yards were located along the canals; those along the Delaware Canal remained dependent on canal boats for deliveries, while those along the Lehigh could have year-round deliveries by train after 1867.

The greatest volume of coal shipped on the Lehigh Canal was 1,276,000 tons in 1855. In September of that year the Lehigh Valley Railroad, built primarily to carry coal, was completed as far as Phillipsburg, New

Jersey, across the Delaware River from Easton.

Competition from the Lehigh Valley Railroad and massive damage to its canals by the flood of 1862 forced the Lehigh Coal and Navigation Company into its decision to extend its Lehigh and Susquehanna Railroad southward. The railroad had been an inclined-plane system connecting White Haven with Wilkes-Barre. The line from White Haven to Mauch Chunk opened in 1866. The Central Railroad of New Jersey leased it soon after its completion to Easton in 1867. The railroad carried trainloads of LC&N coal to markets more distant than the canals could reach.

The decline in the amount of coal carried by boat was rapid once both the Lehigh Valley and Lehigh and Susquehanna (Jersey Central) railroads were in operation. By 1931 it was a mere 65,566 tons.

Coal trains could travel faster and farther than canal boats, but for yards already on the canal it was more convenient and cheaper to continue to have their coal delivered by boat during the season when the canal was open. The yard above became part of the Fritch Coal Company in 1921.

The Fritch Coal Company in Bethlehem, founded in 1921, was the largest coal yard on the Lehigh Canal. It was created from the merger of the Lehigh and Stahr coal yards. *(above)* The front part of a section boat has just been emptied. *(below)* Delivery wagons lined up at the front of the building.

*(above)* Unloading coal from boat 151, captained by William Schaffer, with a bucket conveyer at the Fritch yards in the early 1920s. The hatch covers that covered the coal holds are stacked up behind the man standing in the prow of the boat. *(below)* An uncoupled boat being unloaded at the Saquoit Silk Mill in West Bethlehem, situated between the canal and the river. All the industries along the Lehigh River used anthracite to fuel their steam boilers. The silk industry was introduced into the Lehigh Valley in the early 1880s and became the biggest single industry in the greater Lehigh Valley by the turn of the twentieth century.

*(above)* The steamship Lotta, which carried passengers on the Lehigh River in Bethlehem. July 16, 1886.
*(below)* Lock 41, west of Bethlehem.

# ALLENTOWN

Allentown was already a prosperous small market town and the county seat of Lehigh County when the Lehigh Navigation was completed in 1829. The canal passed on the opposite side of the river from Allentown, but slackwater above the dam at Hamilton Street made both sides of the river navigable. Boatmen poled their boats across the river above the dam from the canal side to Lehigh Port *(seen on page 99)*, an area of the city developed with warehouses and wharfs.

Following the introduction in 1840 of furnaces that used anthracite to smelt iron, the city grew rapidly. Furnaces, foundries, and ancillary manufactories and services of all kinds were started; by the 1850s Allentown had become the largest town in the Lehigh Valley, and its commercial and banking center.

The real growth of Allentown started with the completion of the Lehigh Valley Railroad in 1855. The railroad line passed right through the plants of the Allentown Rolling Mills, the Allentown Iron Company, and the Lehigh Iron Company at Aineyville, which were all built straddling the railroad. After the Panic of 1873 caused a precipitous drop in the demand for iron products, the city's Board of Trade invited silk manufacturers from Paterson, New Jersey, to Allentown to build silk mills. Silk was to become the largest single industry throughout the Lehigh Valley.

*(above)* Loaded boat traveling downstream near the great bend in the river at Allentown. *(left)* Looking upstream toward the city, just above Lock 40.

(above) Loaded boats heading downstream at Allentown.
(right) Driver Lovett Wood with his mules at Lock 40.
(below) Boat 2129 moving downsream past Lock 40,
known as the Dye House Lock because of a nearby silk-
dyeing plant.

*(above)* Saeger's Mill, looking towards the guard lock at the Allentown dam. Mill buildings on both sides of the canal were connected by the covered bridge. *(below)* Looking back at Saeger's Mill from the guard lock. South Mountain is in the background. The sidings and tracks belonged to the Central Railroad of New Jersey, which had a large yard around the bend of the river a short distance downstream.

*(above)* The Hamilton Street bridge and dam, with the guard lock in the foreground. *(below right)* Pleasure boat in Guard Lock No. 7, at the foot of the Allentown dam. This is the same boat that started its trip up the canals on page 9 and is featured on pages 50–51. The view is from north to south. *(below left)* The sluiceway or feeder at the Allentown dam in 1904, with the guard lock and canal on the right. Canal boats entered on the other side of the locktender's house.

*(left)* Work boat in the guard lock above the Allentown dam, heading upstream. On the far left is part of the Allentown Rolling Mills complex, and straight ahead is Adam's Island, formerly known as Haymakers' Island, where a number of local families built summer homes. *(below)* Lehigh Port was on the opposite bank of the river from the canal in Allentown. Boats were poled across the slackwater behind the dam to the many businesses and warehouses there.

Lehigh Canal and River. Allentown, Pa.

214931

*(above)* Postcard view looking south toward Allentown's Hamilton Street bridge, with a canal boat moving upstream in slackwater. Limestone quarries and kilns are on the left, and Adams Island is on the right. The tracks of the Central Railroad of New Jersey are in the foreground. *(below)* Lock 39, now known as Kimmett's Lock, at the end of Section 6 of the canal. The river wall of Kimmett's Lock is preserved and used for a boat launch in the Lehigh River.

*(above)* Work boat and its crew, believed to be in the Catasauqua area. *(below)* Standing is Mrs. Robert Jones (Anna Alabach), seated are Edward J. Albert and Anna T. Geiger, née Albert, on a section boat near the Milson coal yard on Canal Street, below Race Street, Catasauqua.

The Crane Iron Company in the late 1860s or early 1870s *(above)* and in the 1890s *(below)*.

Catasauqua was the site of the first commercially successful use of anthracite to fuel a blast furnace in America, in 1840. The Lehigh Coal and Navigation Company brought David Thomas from Wales to build a furnace here, expecting to vastly increase the market for its anthracite and the use of its canal if he were successful. He was: by 1849 there were five furnaces in Catasauqua.

Anthracite was delivered exclusively via canal to the Crane works until 1868, after the Lehigh and Susquehanna Railroad was completed. The railroad was built by LC&N to compete with the Lehigh Valley Railroad. Catasauqua grew up around the iron works and the canal, becoming a very prosperous town.

David Thomas went on to found an iron company in his own name, but the Thomas Iron Company's many furnaces at many locations used railroads instead of canals for transporting coal and pig iron.

*(above)* Dam and Guard Lock No. 6 above Catasauqua. Across the river is Hokendauqua and the great works of the Thomas Iron Company. Boats entering the canal from the river at this point went past the Crane works a short distance downstream. *(below)* Light boat headed upstream through the guard lock.

(*above*) With a boat in the lock, John Laub, the locktender at Lock 33 at Siegfried, is turning the handle to close the downstream gates before filling the lock chamber. (*below*) Postcard view of Northampton in 1935 showing the Lehigh Valley RR tracks on the left, the Central RR of New Jersey on the right, and the canal winding parallel to the river. The railroad bridge is a single-track connection between the CNJ and the Ironton Railroad, a local line built primarily as an ore carrier and used later by the cement industry. The villages of Siegfried's Bridge (Siegfried), Stemton, and Newport merged to become the Borough of Northampton in 1902.

(above) Standing on a loaded LC&N boat in the outlet lock at Northampton, headed downstream, are, left to right, William Reed, Alex Gold, Joe Reed, Sr., and Louis Possie, the locktender at Lock 35, while Trixie watches the photographer from on shore. The towing post, seen on the left of the boat, was notched so the rope could be pulled or kicked out quickly if necessary. A lantern called a nighthawker was hung from the decorated dasher to help illuminate the way in the dark. (left) Casper Dreher standing by the dasher of the "Evening Star of Weissport," on which would be hung the nighthawker. Not much light was emitted from these lanterns, but it was sufficient for the mules to find their way and for the boat to be visible to oncoming boats.

(*above*) Boats at the loading chute of the Lawrence Portland (later Dragon) Cement Company at Siegfried. This was the only cement company in the Northampton area that loaded cement directly onto canal boats. The other companies shipped their product by rail. The Lehigh Valley was for many years the largest cement producer in the United States, with much of the industry concentrated in the Northampton area. (*below*) "Happy" Joseph Singley and Keiser Rice in the boat basin at Siegfried. In 1923 the canal was closed from Laurys north, and the pay lock was moved downstream to Siegfried. On their return trip from delivering coal, captains of LC&N boats would give the paymaster a slip that recorded the details of the load they had carried, and would then be paid in cash.

Mud digger No. 1 *(above)* outside Lock 32, above Siegfried. The mule stable can be seen near the lock. Maintenance work on the canal system took place during the winter, except for emergency repairs. The mud digger below, at the same location, is unidentified.

*(above)* Rebuilding Slate Dam at Laurys Station in 1902. The Lehigh River valley suffered an extremely destructive flood on February 28, 1902, that damaged dams, canals, locks, railroads, bridges, and numerous buildings. *(below)* The Lehigh Coal and Navigation Company boatyard at Laurys. This became LC&N's main repair and construction yard after 1923, when the Weissport boatyard closed. Laurys became the largest yard ever operated by the company. It closed in 1942.

*(above)* The Laurys boatyard. If boats were kept in good condition, they lasted about twenty years, sometimes longer. Company boatyards were very busy during the winter when maintenance was done. *(right)* The steamboat "Trilby" loading passengers in the canal pool at Laurys. The dams built to water the canal provided popular recreation areas. *(below)* Repairing a disassembled dipper dredge at the Laurys boatyard are, left to right, George "Pudding" Reed, Harvey Dotter, Paul Fenstermaker, Fred Walck, and Clayton Dotter.

Coal accumulated behind the dams in such great quantities that it became financially worthwhile to reclaim coal from the silt. A reclamation plant was located at Treichler's, seen here in 1918. Other reclamation operations took place at Laurys Station, Three-Mile Dam, Lehigh Gap Dam, and at the Glendon level.

By the early 1920s no coal was shipped by canal out of Mauch Chunk. Instead, it was loaded onto trains and transferred to canal boats at a transfer plant at Slate Dam (Laurys). Boat 296, operated by John Minder, was the first boat to be loaded, on June 11, 1923. On the left *(below)* is George Searfoss. His father John is the man on the right wearing suspenders.

(*above*) A boat carrying coal dirt in Lock 27 at Lockport, headed upstream with its hatches off.
(*below*) Dam and Guard Lock No. 4 at Treichler's.

*(left)* Dipper dredge below Walnutport.

*(below)* Lock 23, south of Walnutport. The building on the right was in use as a garage when this photograph was taken. It is the same building *(bottom)* that had been used as a lockside store where boatmen could buy provisions. This lock was also known as Lock 22. When the canal was re-engineered to remove one lock and deepen another, lock numbers could change officially, but users of the canal did not always respect the formal change.

*(above)* Lock 23, Kelchner's Lock, south of Walnutport, in the early twentieth century. Frank Kelchner was the locktender. *(below)* The canal was used, on Sundays particularly, for recreation. This group is enjoying an outing along the canal at Walnutport.

*(above)* Unloading coal at Wilson Kelchner's coal yard in Walnutport, using a derrick, block and tackle, and draft horses to hoist buckets of coal from the boat. Compare this method with the much more efficient system used elsewhere, such as at Leedom's coal yard at Yardley and the Fritch Coal Company in Bethlehem. Smaller yards did not use mechanized equipment; instead, men shoveled coal from the holds. Boatmen could sometimes pick up a few extra dollars doing this unpopular job themselves. The man with the moustache is Harry E. Kelchner. *(below)* One of the Seitz beer boats moored at the Anchor Hotel at Walnutport. This section of the canal was called the Anchor level because of the importance boatmen attached to this hotel.

*(above)* Looking north into Lehigh Gap. The old highway bridge, Chain Bridge, can be seen in front of the Lehigh and New England railroad bridge. Gap Dam, Dam No. 3, is in the foreground.

*(left)* Coal train seen from the Lehigh and New England railroad bridge at the Gap. From the 1860s trains carried more coal than canal boats.

(left) Tug boat and scow loaded with dredged material at the Reber Brothers facility in Bowmanstown at Dam No. 2. (below) Closeup of the coal dredge used to reclaim coal from behind the dam. The worker on the right is Maurice Bomba. (bottom) Dredge at work behind Lehigh Gap Dam.

CANAL LOCK BELOW LEHIGH GAP.
SLATINGTON, PA.

(left) Lock 21, below Lehigh Gap. (below) The aqueduct over the Aquashicola Creek at Lehigh Gap. (bottom) Lock 18, above Lehigh Gap, in 1886.

*(above)* Looking upstream from Lehigh Gap toward Palmerton. The large boat basin allowed boats to turn easily, and was necessary to store sufficient water for the lock since there was not a long level to draw from. *(below)* LC&N and private boats docked in front of New Jersey Zinc's spiegeleisen furnaces at Palmerton, in which ferro-manganese was produced from zinc residuum. The ferro-manganese was used in Bessemer converters and open-hearth furnaces to produce steel. New Jersey Zinc was the largest consumer of coal reclaimed from silt. Canal transportation of coal silt stopped in 1942. Dredging behind the dams continued until about 1960, with the coal silt taken to Palmerton by truck.

*(above)* A geared-down Fordson tractor owned by one of the coal-dirt reclamation companies hauling a loaded boat above Bowmanstown. *(below)* Heading upstream through the lock at Parryville, Lock 13.

(above) Dredge filling a canal boat with coal silt above Bowmanstown. In the background is the Carbon Iron Company furnace at Parryville. (below) The Parryville Furnace of the Carbon Iron and Pipe Company, about 1900. The furnace was built originally in 1855, and for its first years depended on the canal for raw products and shipping its pig iron. The locktender's house for Lock 13 is barely visible in front of the furnace.

*(above)* The old bridge at Weissport in 1886, photographed by Louis Tiffany. *(below)* The Lehigh Coal and Navigation Company's Weissport boatyard. The buildings are, left to right, lumber storage shed, feed house, and parts storage shed.

(*above*) Louis Tiffany photograph of youngsters, the girls in sunbonnets, sitting on a boat at the Weissport boat yard. The Molly-Polly-Chunker went through Weissport on a Thursday, yet the people on the boat appear to be dressed in their Sunday clothing. (*below*) Packerton, looking down on the big Lehigh Valley Railroad yards and shops on the west bank of the Lehigh River. The canal is on the east side of the river. The Central Railroad of New Jersey is on the same side of the river as the Lehigh Valley Railroad through this section. The building on the top of the hill is the local school.

Photographs by Louis Tiffany and Walter Tuckerman of the locktender at Lock 4, just south of Mauch Chunk, *(right)* sitting on the stern of a section boat and *(below)* standing in his tiny patch between the canal and the mountainside. A sparsely furnished cottage, the lock house, a rubbish-filled stable yard and a small, stony vegetable garden comprised the family's holdings. Louis Tiffany is on the right, taking a photograph of the locktender. Standing next to him is Louise Knox, his future wife.

(*left*) The locktender feeding some of his eight goats, which provided milk for the family. (*below*) The log of the Molly-Polly-Chunker describes the wife of the lockkeeper at Lock 4 as a fine-looking woman, the mother of eight children, all under twelve, six of whom are in the photo below. None of them could read, or speak more than a few words of English. The mother did some wash for the travelers, which they picked up on their return trip. Most people along the canal were bilingual, but Pennsylvania German was the common language of the canal, especially from Allentown north — even black and Irish boatmen spoke it.

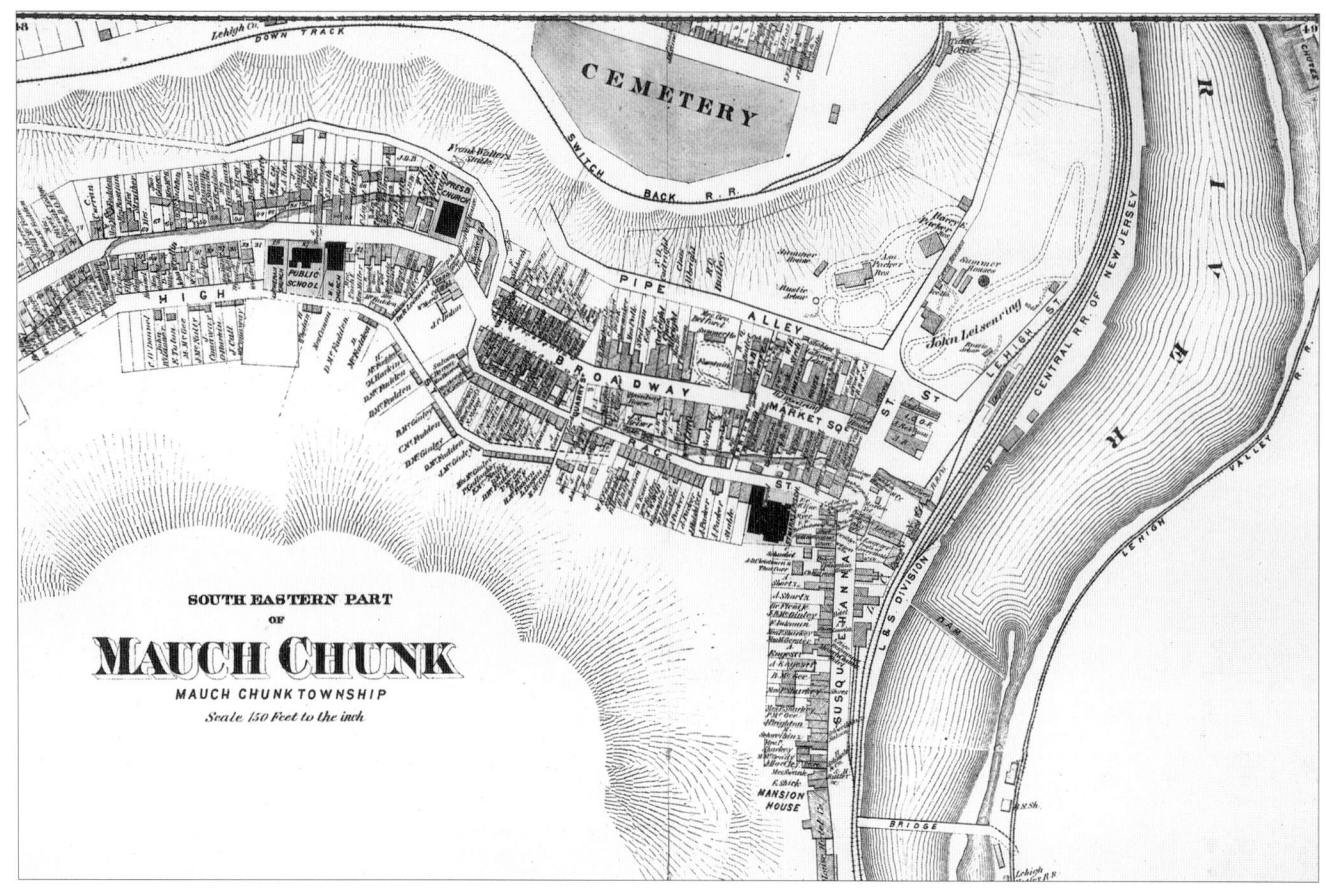

## MAUCH CHUNK

Mauch Chunk developed as a synthesis of coal, the Lehigh Canal, and the Lehigh Valley and New Jersey Central railroads. During the nineteenth century, when tens of millions of tons of anthracite were shipped out from Mauch Chunk on the canal and the railroads, almost everyone in the town was employed directly or indirectly by coal or coal-transportation interests. With the completion in 1829 of the Lehigh Canal, in 1855 of the Lehigh Valley Railroad, and in 1867 of the Lehigh and Susquehanna Railroad (which was leased in 1871 to the Central Railroad of New Jersey) Mauch Chunk became the center of coal transportation for the great anthracite coal fields of northeast Pennsylvania. There was constant movement of canal boats and coal trains through the town.

Coal and railroad magnates constructed mansions in Mauch Chunk, but left when the Lehigh Valley Railroad moved its operations headquarters to South Bethlehem in 1857. During the 1870s a complete reorganization of the Lehigh Coal and Navigation Company and its takeover by outside investors moved the headquarters of the company from Mauch Chunk to Lansford, where a large modern office building was built in 1873.

During the mid- and late nineteenth century the scenic splendor of the Lehigh River gorge and attractions such as the Switchback Railroad made the Mauch Chunk area a popular resort known as the "Switzerland of America," favored by wealthy city people seeking to escape the sweltering heat of summer.

The coal chutes, where coal was loaded into canal boats or railroad cars, were above Mauch Chunk just beyond the top of the 1870s map above. The weigh lock, where boats were weighed before heading down the canal, was a short distance downstream of the Lehigh Valley depot, slightly below the map.

Stereopticon view of Mauch Chunk in the 1870s or early 1880s, showing much of what can be seen in the map opposite.

(*above*) Accumulation of boats between the weigh lock and Lock 2. The weigh lock is the structure between the two basins. The bridge downstream is where, because of topographical difficulties, the Lehigh Valley Railroad crossed over to the right bank and ran parallel to the tracks of the Central Railroad of New Jersey as far as Lehighton, where the Central crossed to Weissport on the east bank and continued down the river on the left bank. (*below*) Rebuilding the water intake for the Mauch Chunk hydroelectric station in the weigh lock level in March, 1909.

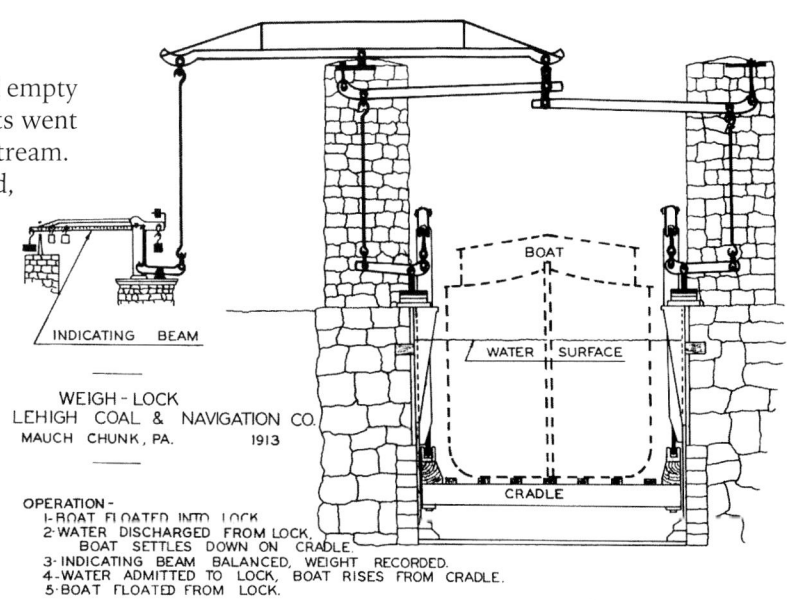

*(above)* The weigh lock in the late teens or early twenties of the twentieth century. On the far right is Ervin Ahner; on the far left is Harry E. Solt. Little is known about the operation of the weigh lock because no written record remains. It was not used after 1922.

## WEIGH LOCK

All company and private boats were weighed empty at the beginning of each season, and all boats went through the weigh lock on their way downstream. Private boats paid a toll after being weighed, company boats were simply logged. Boats carried 90 to 100 tons of coal, and each captain of a company boat was allowed an additional 200 pounds per trip for his own use — to store for his own use during the winter, to sell, or to trade.

INDICATING BEAM

BOAT

WATER SURFACE

CRADLE

WEIGH-LOCK
LEHIGH COAL & NAVIGATION CO.
MAUCH CHUNK, PA.        1913

OPERATION-
1-BOAT FLOATED INTO LOCK.
2-WATER DISCHARGED FROM LOCK,
    BOAT SETTLES DOWN ON CRADLE.
3-INDICATING BEAM BALANCED, WEIGHT RECORDED.
4-WATER ADMITTED TO LOCK, BOAT RISES FROM CRADLE.
5-BOAT FLOATED FROM LOCK.

TRACED FROM LC.& N. CO. DRAWING
AUG. 1963   J.R.CONNELLY

Mules were dependable and strong. They could work 16-hour days, resting only when their boat was going through a lock; but unlike horses, mules would not work themselves to death. They were capable of pulling a loaded boat at a steady pace of two to three miles per hour, and could make the round trip from Mauch Chunk to Bristol and back, 212 miles, in seven or eight days. Their drivers were often boys who went on to captain their own boat or even to own one or more boats. *(above)* Left, Ted Sherman of Weissport and Lucy; right, Frank Wells of Uhlertown and Dan at Lock 3, below Mauch Chunk, on July 3, 1913. *(below)* John S. Grey and mules at Lock 3 in 1916.

*(above)* Daniel McCollick of Erwinna with his team at Mauch Chunk in 1911. *(below)* An unidentified driver with his team at the Mauch Chunk loading docks . The mules are wearing fly netting and decorated harness.

View from Prospect Rock, looking upstream toward the dam and guard lock on the right. Many boats can be seen in the pool at the foot of Mount Pisgah, waiting to be filled with coal at the chutes. The Beaver Meadow coal chutes at East Mauch Chunk are barely visible across the river and a little upstream. The Beaver Meadow Railroad was the earliest railroad down the Lehigh River valley, dating from 1836. Coal mined at Beaver Meadows, south of Hazleton, was transferred here to canal boats for shipment south. The Lehigh Valley Railroad grade can be seen on the right. The bridge in the foreground connected the Lehigh Valley depot with the main part of town.

(above) Early view of Mauch Chunk from Prospect Rock. The station of the Central Railroad of New Jersey has not yet been built, but the tracks can be seen on the left. Two buildings seen here, the Carbon County Democrat and the Mauch Chunk Steam Printing House, were torn down in 1888 to make way for the station. The tracks of the Lehigh Valley Railroad are on the opposite side of the river.

(right) View from Prospect Rock, looking downstream toward the Mauch Chunk Narrows. The log of the Molly-Polly-Chunker describes a night of fitful sleep here because of "the vigorous salutes of passing freight trains ... Our way at this point lay between two mountains and was shared by two lively but inartistic railroads, one on each side of the Lehigh River."

*(above)* Mauch Chunk viewed from the south. The town was extremely restricted because of the steep hillsides and the narrow valley of the Lehigh River, and was unable to continue growing. *(below)* A Seitz beer boat docked in front of the Lehigh Valley depot.

(above) In the foreground is the guard lock; across the river is the Mansion House on Susquehanna Street. Built in the 1830s and torn down in 1914, it was the most prominent hotel in town. While it was in Mauch Chunk (below, with Miss Knox in the bow) the Molly-Polly-Chunker was visited by the wife of the proprietor as well as several well-dressed city people. It was a source of great curiosity, as much for the boat itself as for the passengers, whose arrival had been announced in advance by the local newspaper: "The boat used ... is a handsome barge conveniently arranged into six different apartments, consisting of dining-room, parlor, sleeping apartments, kitchen, etc. The inside decorations consist of Japanese designs, lanterns and bric-a-brac generally; the sitting room is well fitted out with books, maps of the different counties in the states through which the party is to pass, photographic apparatus, etc. The floors of the different rooms are carpeted, and the culinary department is presided over by two colored servants."

Summer and winter views of Lock No. 1 and Dam No. 1. The Lehigh Valley depot is on the far right, and the passenger station of the Central Railroad of New Jersey can be recognized by its conical roof. The photograph above was possibly taken on July 4, around 1900, as all the boats are tied up for a holiday.

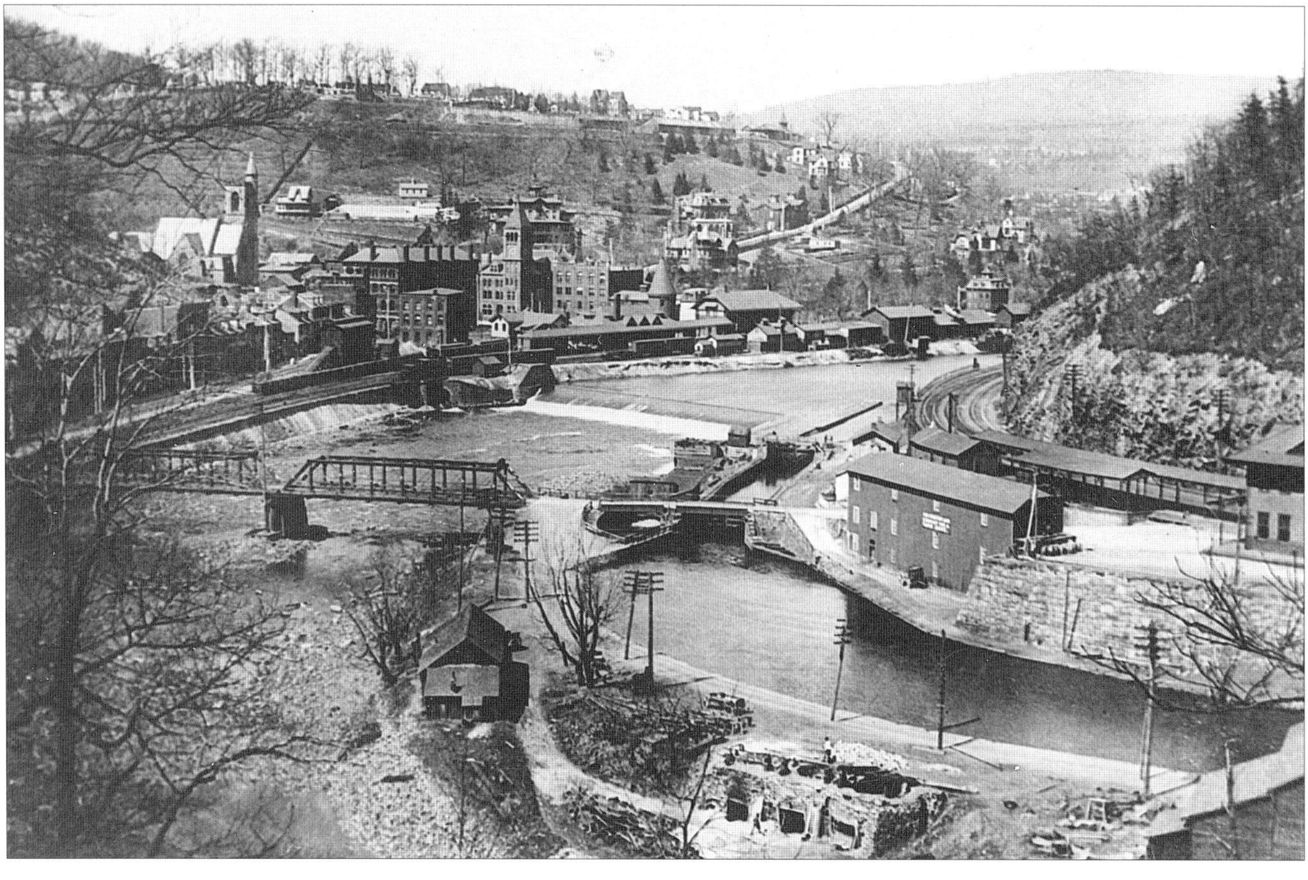

Lock No. 1 of the Lower Division. *(above)* So much water is being diverted to feed the canal that there is very little flowing in the river bed. *(below)* The lock chamber of Lock 1, which was more than twice the length of normal locks.

The flood of December 15, 1901, caused much damage to the Lehigh and Delaware canals. A much worse one came on February 28, 1902 (March 2 on the Delaware), resulting in such extensive damage that no coal was shipped for two years. *(above)* Spectators lined up at the Central station to watch the rising waters on December 15, 1901. *(below)* Tracks of the Central Railroad of New Jersey undermined by rampaging floodwaters.

*(above)* View from the towpath of the damage left by the December 1901 flood. *(below)* The canal choked with ice and debris below the Lehigh Valley station, probably after the February 1902 flood.

*(above)* View across the river of the coal chutes at the foot of Mount Pisgah, before 1873. In the foreground LC&N cars are lined up on Lehigh Valley Railroad tracks on the site of the former Beaver Meadow coal chutes, which were removed after the takeover of the Beaver Meadow system by the Lehigh Valley in 1864. *(below left)* Overview from Mount Pisgah of chute No. 1, at the foot of the inclined plane, showing the connecting line from the switchback to the loading area, with many section boats waiting to be filled. *(below right)* The bottom of chute No. 1 in the late 1860s.

(above left) Stereopticon view from the river up the coal chutes and inclined plane at Mount Pisgah. (above right) Section boats waiting to be filled at the Beaver Meadow loading docks. (right) Boats being loaded at the foot of Mount Pisgah. The boats in the foreground look like Morris Canal boats. While LC&N boats rarely traversed the Morris Canal, and could not do so unless they were carrying less than 75 tons, Morris Canal boats traveled regularly on the Lehigh Navigation, crossing the Delaware River between Phillipsburg and Easton. (below) Loading pens for railroad cars on the Central Railroad of New Jersey, at the foot of the gravity railroad.

## THE SWITCHBACK RAILROAD

The Lehigh Coal and Navigation Company built a gravity railroad from Summit Hill to Mauch Chunk in 1827 to carry coal from their mines to the shipping point on the canal. Until the backtrack with its inclined planes was built in 1843–1844, empty cars were pulled back to the mines by mules, up the gravity railroad. After the completion of the Hauto tunnel in 1872 coal was carried through the mountains by train; the planes and gravity railroad, from their earliest days a tourist attraction, were then used exclusively as the famous "Switchback Railroad" that took tourists on a roller-coaster type of ride across the hills north and west of Mauch Chunk. *(left)* Tourist car at the foot of the Mount Pisgah plane, ready to be rolled forward and switched to the other track where it will be attached to the barney waiting in the pit. *(above)* The barney that pushed cars up Mount Pisgah is on the left. The plane had an elevation of 664 feet and was 2,322 feet long. From the summit cars rolled down a gravity track, dropping 302 feet in almost seven miles to the Mount Jefferson plane *(above left, with coal jimmies waiting to be attached to the barney)*, which raised them 462 feet in 2,070 feet. From Mount Jefferson they descended 45 feet in one mile to the mining town of Summit Hill. The return trip from Summit Hill to Mauch Chunk was entirely by gravity. The round-trip distance was 18 miles.

Massive, 30-foot-diameter steam-driven gears at the top of the Mt. Pisgah plane pulled the barney to the top by winding two wrought-iron bands attached to the barney around a huge drum. *(right)* A tourist car being pushed up Mt. Pisgah by a barney, which was being pulled up the plane by 7½-inch-wide iron straps. The straps on each track were attached to a wire rope that traveled under the barney pits and kept the system taut. *(below)* Repairs underway at the Mt. Jefferson plane.

Packer Dam, Dam No. 1 of the Upper Grand Section of the Lehigh Navigation, was at East Mauch Chunk, a short distance upstream from Mauch Chunk. Across the river are the shops and yards of the Central Railroad of New Jersey. Flash boards can be seen on the top of the dam. These raised the level of the river by six inches or more.

## THE UPPER GRAND SECTION

The Upper Division of the Lehigh Navigation system was started in 1835 and opened in June of 1838. An engineering marvel, featuring some of the largest locks constructed anywhere during the canal era, the Upper Grand Section dropped 600 feet in the 26 miles from White Haven to Mauch Chunk. Of the 20 dams, Dam 8 at Barn Door was 38 feet high and only two were under 20 feet in height. All the locks were deep. There were 29 in the Upper Division; of these, 26 were deeper than 15 feet, and 22 were at least 20 feet deep.

While the Upper Division was being built, the Beaver Meadow Railroad was constructing a line parallel to the waterway, leading to frequent conflict between construction crews. Whenever a crew began work on a section of the railroad, an LC&N crew would appear to clear timber for canal locks. By coincidence, LC&N contended, its crew would happen to be directly above the railroad workers and would tend to send logs dropping into the competitor's work site. Fights ensued, and the Beaver Meadow Railroad had a friendly law enforcement officer form a posse to arrest the LC&N foreman. Eventually, the court favored LC&N

because of its state-chartered near-monopoly rights to the corridor. The companies were directed to exchange information on their work schedules, but LC&N plans received priority.

The Upper Division offered a spectacular trip by boat through the Lehigh Gorge. At White Haven the transportation system continued to Wilkes-Barre by the LC&N's Lehigh and Susquehanna Railroad because of the barrier of the mountains.

Trade on the Upper Grand Section grew and prospered through the 1840s and 1850s. In 1862, however, the Lehigh River valley was devastated by a flood. Small mill dams started breaking during a heavy rain; the force of this water caused log booms to break, sending hundreds of thousands of logs crashing through the large dams, releasing enormous volumes of water and smashing everything in their path. The flood destroyed the Upper Division and did great damage to the main navigation downstream of Mauch Chunk. The Upper Division was never restored for navigation above Coalport. It was replaced by an extension of the Lehigh and Susquehanna Railroad.

The Lehigh Coal and Navigation Company built extensive coal-loading docks at Coalport, one mile above Mauch Chunk, in 1872–73. Anthracite was brought here by trains to be loaded onto canal boats. *(top)* The Coalport loading pockets about 1885; *(left)* about 1900; *(above)* in 1907. The cars in the top picture are Central Railroad of New Jersey cars. The tracks crossed the river a short distance upstream to enter Mauch Chunk.

(*above*) The Molly-Polly-Chunker entering Lock 1 of the Upper Division at Packer Dam, below Coalport. The log of the journey described this as a place where there were "many boats, and more flies." While the boat was there, several of the party went on an excursion to nearby Glen Onoko by railroad. Coalport was the end of the navigation after 1862, and the boat turned around here to return to Bristol. (*below*) Lehigh Coal and Navigation Company boats and private boats moored in the basin upstream of the loading docks at Coalport early in the twentieth century. The reason for so many idle boats is unknown, but it may have been during the "Great Strike" of 1902, when anthracite miners were on strike from the beginning of June to the end of October. The view is looking north.

The log of the Molly-Polly Chunker records that at Glen Onoko the travelers "found hurdy-gurdies, brass bands, a large hotel, miniature coal-breakers, boys selling rhododendrons, and a beautiful ravine decked out with American flags." *(above)* The Hotel Wahnetah at Glen Onoko could be reached only by train. It was a popular resort in the "Switzerland of America," where city people would come to enjoy the cool summer air. The paths, bridges and stairways up the glen have long since gone back to nature, and the hotel burned down in 1917. *(below)* Upper Division Lock No. 1 at Packer Dam in later years. The dam can be seen on the left. In the background are the shops of the Central Railroad of New Jersey.

*(above)* The Penn Haven planes. The Beaver Meadow planes are on the left, completed in early 1851; on the right are the planes of the Hazleton Coal Company, opened later the same year to carry its coal to the Lehigh River to be transferred to canal boats. After the completion of the Lehigh Valley Railroad in 1855 some Hazleton coal was shipped from Penn Haven on the Beaver Meadow Railroad to East Mauch Chunk, and from there on the Lehigh Valley Railroad. *(below)* 1860 lithograph of the coal-loading docks at Penn Haven, which were destroyed in the 1862 flood.

The dam at Tannery, 34 feet high, was destroyed in the 1862 flood and never rebuilt. The Upper Grand Section was a major tourist attraction before its destruction. Americans and Europeans traveled to Mauch Chunk to view the engineering marvels of the area — both the Upper Grand Section and the Mauch Chunk Gravity Railroad, which later became known as the Switchback.

*(above)* Lock 24 at Dam No. 16 of the Upper Grand Section at Tannery, south of White Haven. This lock had a lift of 30 feet and was known as "the Pennsylvania Lock." It is now a pile of stones. *(below)* The canal corridor is now part of Lehigh Canal State Park. Lock 28 in the upper division is one of several locks where volunteers have cleared out trees and undergrowth so the original size can be seen. An observation platform has been placed on the top so visitors can look down into the 22-foot-deep lock.

Two dams in the upper section, at White Haven and at Bridgeport, were rebuilt and maintained after the 1862 and subsequent floods because of many water-powered mills and industries that depended on them. *(above)* The second Lehigh Valley Railroad bridge at Bridgeport, just south of White Haven, was completed in 1887. It replaced the original 1867 wooden bridge. *(below)* It was common practice to place railroad cars on bridges to provide extra weight to try to prevent their collapse during floods. The new LVRR bridge, completed in 1901, was destroyed in the flood of 1902.

Repairing the dam at White Haven in the summer of 1908. *(above)* Filling the empty cribs under the apron with stone. *(below)* The scow used to carry stone to the dam, floating here behind the dam, was loaded onto a flat car at Parryville and shipped by rail to White Haven.

View toward the south at the base of plane No. 2 of the Ashley Planes in about 1930. About 3,000 feet long, with a slope of 14.65%, it was built in 1865–67, replacing the original 1847 No. 2 plane. A hoisting engine with a drum 20 feet in diameter and a steel cable 2½ inches in diameter are pulling the barney up the plane.

## THE ASHLEY PLANES

The Ashley Planes mark the end of the Lehigh Coal and Navigation Company's system of water and rail transportation that linked the Susquehanna and Delaware rivers and carried coal from the anthracite fields of Pennsylvania to Philadelphia and New York, and markets between and beyond. They were constructed because, under the provisions of its charter from the state, the Lehigh Coal and Navigation Company had to establish a transportation link with the Susquehanna River near Wilkes-Barre.

To resolve the formidable difficulties in building a canal from the northern terminus of the Upper Grand Section of the Lehigh Navigation at White Haven over the 2,000-foot-high Penobscot Mountain, the company elected to construct the Lehigh and Susquehanna Railroad. The building of this railroad was a remarkable achievement, involving the construction of a 1,743-foot tunnel between White Haven and Solomon Gap and a series of three double-tracked inclined planes from Solomon Gap to Ashley near the Susquehanna River. By 1843 the railroad was in operation. Improvements were made during 1847–48, when steam engines and wire-rope cables were installed on the planes.

Although in early years a few canal boats passed over the planes, the primary traffic was railroad cars carrying cargoes of Wyoming Valley anthracite to canal boats at White Haven.

## EXPERIMENTS

In 1907 LC&N tried a number of innovative ways to modernize canal transportation on the canal, hoping to cut costs and save time. The steel canal boat *(upper left)*, built in Philadelphia, was too heavy at 20 tons unloaded, and would not respond to rudder. It made only one trip *(right)* carrying a load of coal. Photos were taken above Packer Dam. Early in 1907 narrow-gauge tracks were laid for a mine locomotive to pull boats from Coalport three miles to the weigh lock; in September an "electric mule" on a monorail operated briefly between the weigh lock and Lock 7 at Weissport, again a distance of three miles. Both these methods proved too costly. More electricity was used than anticipated, and the increased speed of the boats caused so much wash that the banks were damaged. In 1909, a specially adapted 60-horsepower motor truck hauled four loaded boats from Allentown to Bethlehem. If this method had been adopted, the towpath would have had to be extensively improved. The mule team prevailed for as long as boats plied the canals.

*(top)* Entrance to the lock at the Coalport level. *(above)* Entering the lock at Packer Dam, with the mine locomotive on the towpath. *(left)* The locomotive and its trolley.

Two motors mounted on a monorail were built for the Lehigh Coal and Navigation Company by the General Electric Company. They required two men to operate them, a motorman and a controller. Electricity was provided by the Mauch Chunk Electric Light Company. The cars weighed three tons and ran on an "I"-beam track.

*(above)* The inventor of the electric mule, Leon Girard, and his son Ernest. *(below)* A four-boat tow going up the canal, pulled by the electric mule. Four empty or two loaded boats could be pulled by this method. The shops and yards of the Lehigh Valley Railroad at Packerton can be seen in the distance.

# PHOTO CREDITS

With the exception of those listed below, all photographs used in this book are from the collections of the National Canal Museum in Easton, Pennsylvania. Most of these are in the Pennsylania Canal Society collection, which is held at the canal museum archives.

From the collections of the Lehigh County Historical Society:

Workboat photos on page 99 and 101

Crane Iron Company in the 1890s on page 102

Closing Lock 33 at Siegfried on page 104

Lockside store on page 113

From the Raymond Holland Collection:

Lehigh Port in Allentown on page 99

Group of young men and women on page 114

From Wildlands Conservancy:

Lock 28 in Lehigh Canal State Park, photo by Tom Gettings, on page 150

# BIBLIOGRAPHY

*Agreements Between the Lehigh Canal and Navigation Company and the Easton Power Company and Report on the Glendon Level.* Easton, 1901.

Anderson, John A. "Navigation on the Delaware and Lehigh Rivers." In *Proceedings of the Bucks County Historical Society*, Vol. IV, 1917: 282–312.

Barber, David. *A Towpath Guide to the Lehigh Canal, Lower Division.* Philadelphia: Delaware Valley Appalachian Mountain Club, 1983.

Bartholomew, Craig L., and Lance E. Metz. *The Anthracite Iron Industry of the Lehigh Valley.* Easton: Center for Canal History and Technology, 1988.

Bryski, Anthony J. "The Lehigh Canal and its Effects on the Economic Development of the Region through which it Passed." Ph.D. dissertation, New York University, 1957.

Campion, Joan. *Smokestacks and Black Diamonds: A History of Carbon County, Pennsylvania.* Easton: Canal History and Technology Press, 1997.

Coleman, Lyman. *Guidebook of the Lehigh Valley Railroad.* Philadelphia: J.B. Lipincott & Co., 1872.

*Delaware Canal Master Plan.* Friends of the Delaware Canal, 1987.

Eschenbach, G.W. *The Forks of the Delaware 1752–1900.* Easton: Eschenbach Printing House, 1900.

Fackenthal, B.F. "Improving Navigation on the Delaware River with Some Account of Ferries, Bridges, Canals and Floods." In *Proceedings of the Bucks County Historical Society*, Vol. VI, 1932: 103–320.

Gilbert, Joan. *Gateway to the Coalfields.* Easton: Canal History and Technology Press, 2005.

Henry, M.S. *History of the Lehigh Valley.* Easton: Bixler & Corwin, 1860.

Hoffman, Phillip. *Anthracite from the Lehigh Valley Region of Pennsylvania 1820–1845.* Washington, DC: Smithsonian Institution, 1968.

Hydro, Vincent, Jr. *The Mauch Chunk Switchback, America's Pioneer Railroad.* Easton: Canal History and Technology Press, 2002.

Klein, Theodore B. *The Canals of Pennsylvania.* Harrisburg, 1901.

Knies, Michael. *Coal on the Lehigh 1790–1827: Beginnings and Growth of the Anthracite Industry in Carbon County, Pennsylvania.* Easton: Canal History and Technology Press, 2001.

_____. "Industry, Enterprise, Wealth and Taste: The Development of Mauch Chunk." In *Canal History and Technology Proceedings* Vol. IV, 1985: 17–45.

Kulp, Randolph L. , ed. *Railroads in the Lehigh River Valley.* Allentown: National Railway Historical Society, 1974.

Lee, James. *The Morris Canal: A Photographic History.* Easton: Delaware Press, 1979.

*Lehigh Canal: An H.C.R.S. Project Report.* Washington: U.S. Department of the Interior, Heritage, Conservation, Recreation Service, 1980.

Log of the Molly-Polly-Chunker.

McClellan, Robert J. *The Delaware Canal, A Picture Story.* New Brunswick: Rutgers University Press, 1967.

McKelvey, William J. *The Delaware and Raritan Canal: A Pictorial History.* York, PA: Canal Press, Inc., 1975.

Metz, Lance E., and Donald Sayenga. *Capt' Sherman's Guide to Hugh Moore Park.* Easton: Center for Canal History and Technology, 1988, reprint 1998.

Miller, Benjamin LeRoy. *Pennsylvania Geological Survey: Lehigh County.* Harrisburg, 1941.

Miller, Benjamin LeRoy, Donald McCoy Fraser and Ralph LeRoy Miller. *Pennsylvania Geological Survey: Northampton County.* Harrisburg, 1939.

Miller, John P. *The Lehigh Canal, A Thumb Nail History 1829–1931.* Bethlehem, 1979.

Morton, Eleanor. *Josiah White, Prince of Pioneers.* New York: Stephen Daye Press, 1946.

Parton, W. Julian. *The Death of a Great Company.* Easton: Center for Canal History and Technology, 1986, reprint 1998.

Rivinus, Willis M. *A Wayfarer's Guide to the Delaware Canal.* Doylestown, PA: 1964.

Russo, A.L. "A Historical Survey with Maps of Industrial Sites along the Lehigh Canal 1830–1880." Unpublished research paper. Lafayette College, Easton, 1980.

Sayenga, Donald. "The Mauch Chunk Wire Rope Factory." In *Canal History and Technology Proceedings* Vol. XVII, 1998: 135–158.

Sears, John F. *Tourists in an Industrial Scene, Mauch Chunk, Pennsylvania.* Jim Thorpe, PA: Mauch Chunk Historical Society, 1983.

Taylor, Frank H. *Autumn Leaves Upon the Lehigh.* Philadelphia: James W. Nagle, 1876.

Waltman, Charles. "The Influence of the Lehigh Canal on the Industrial and Urban Development of the Lehigh Valley." In *Canal History and Technology Proceedings,* Vol. 11, 1983: 87–104.

Williams, David L. "The Lehigh Canal System, The Lehigh Coal and Navigation Company." In *Proceedings of the Lehigh County Historical Society,* Vol. XXII, 1958: 97–137.

Young, W.S. *History of the Lehigh Coal and Navigation Company.* Philadelphia: Lehigh Coal and Navigation Company, 1840.

Yoder, C.P. "Bill." *Delaware Canal Journal.* Bethlehem: Canal Press Incorporated, 1972.

Zimmerman, Albright G. *Pennsylvania's Delaware Division Canal: Sixty Miles of Euphoria and Frustration.* Easton: Canal History and Technology Press, 2002.

# INDEX

## A

Abbott Street, 72, 75

agriculture, 56

Ahner, Ervin, 129

Albert, Edward, 101

Allentown, 95–100

Allentown dam, 98

Anchor Hotel, 115

anthracite, 4–5, 91
    use of by industries, 93
    use of for ironmaking, 102

Aquashicola Creek aqueduct, 118

aqueducts, 44, 48, 61, 118

Ashley Planes, 153

## B

Bagg, Lyman, 70

Beaver Meadow, 132
    chutes, 140
    planes, 148
    Railroad, 132, 144

beer boat, 115, 134

Best, Captain John and Martha, 70

Best, Robert and Martha, 43

Bethlehem, 86–94

Bethlehem Steel, 86

boat basins, 11, 32, 67–68, 75, 119, 146

boat capacity, 5, 66, 129

boat dimensions, 33, 54, 68, 76

boat maintenance, 109

boat tows, 9

boat yards, 52, 55

boatmen's supplies, 85

boats on restored canal, 77

boats passing, 32, 43

boats under construction, 55

boatyards, 108–109, 122

Bowmanstown, 117, 120

Boy Scouts, 45

Bridgeport, 151

bridges, 15, 24, 38, 88, 104
    covered, 53, 57, 77, 97, 114, 115, 121
    Mauch Chunk, 132, 133, 135

bridges
    pedestrian, 72
    railroad, 64, 66–67, 72, 73, 74, 151

Bridgeton, 55, 56

Bristol, 9–17

Brown, Jimmy, 47

## C

cable ferry, 27, 68

camelback bridges, 20, 21, 24, 50, 51

cameras on Molly-Polly-Chunker, 56

Canal Museum, National, 3, 77, 78, 80

canals, 3, 4–5
    dimensions of, 5, 69, 144
    length of, 5, 66, 69
    Delaware, 7–68
    Delaware and Raritan, 27–28
    Lehigh, 69–157
    Morris, 66–68
    Upper Grand Section, 144, 146, 148–153

Carbon Iron and Pipe Co., 121

Catasauqua, 101–103

cement industry, 106

Central RR of New Jersey, 73, 87, 97, 104, 126, 127, 133, 141
    shops of, 144, 147

Chain Dam, 78–79, 80, 81

Change Bridge, 80

children photographed by Tiffany, 24, 55, 56, 59, 65, 123, 125

chunkers, 35

Church, Superintendent I.M., 45

coal chutes, 132, 140–141

coal reclamation, 119, 120

coal silt, 110–112, 117

coal train, 116

coal unloading, 12, 22, 42, 91–93, 115

coal yards, 11, 12, 22, 39, 91–93, 101, 115

coalfields, map of, 2

Coalport, 145–147, 154–155

competition from railroads, 91

Cosman, John, 19

Coyle, Vickie, 77

Crane Iron Co., 102

crossing river, 80

## D

dams, 69, 71, 87, 98
    rebuilding of, 108, 151, 152

Davies-Strauss Stauffer Co., 90

deep locks, 144, 150

Delaware and Raritan Canal, 27–28

Delaware Canal State Park, 7, 54

Delaware Canal, history of, 7

Delaware River, 9, 38–40, 66–67, 69

dimensions of boats, 54, 76

Morris Canal, 66–68

doghouse, 13, 24

Dotter, Harry and Clayton, 64, 109

dredging, 80, 107, 110, 113, 117, 119, 121

Dreher, Casper, 105

drivers, mule, 19, 84, 130–131

Durham, 60–61

Dye House Lock, 96

## E

Easton, 65, 69–75
    dam, 71
    guard lock, 64, 65

Easton and Northern RR, 74

electric mules, 154–157

Emery, Captain Grant, 21

Erwinna, 49, 50

experimental boat hauling, 154–157

## F

family boats, 21

Fenstermaker, Paul, 109

filling in canals, 58

floods, 39–41, 44, 62, 91, 108, 144, 148, 149, 151
    at Mauch Chunk, 138–139

fly netting, 131

Forks of the Delaware, 69, 71

Frank, Sam, 69

Freemansburg, 84–86

Friends of the Delaware Canal, 7

Fritch Coal Co., 87, 92–93

furnaces, iron, 60, 76

Palisades, 57–58

Palmerton, 119

Parryville, 120

passing boats, 32, 43

pay lock, 106

payment of LC&N boat captains, 106

Penn Haven planes, 147

Phillipsburg, 66–67

Pisgah, Mt., 140–143

pleasure boating, 9, 18, 36, 45, 50–51, 57, 77, 94, 98, 109

Point Pleasant, 42–43

Port Delaware, 64, 66–68

Possie, Louis, 105

private boats, 15, 42

**Q**

quarries, 38, 100

**R**

Rabbit Run, 37

railroad bridges, 64, 66–67, 72, 73, 74

railroads, 87, 126, 132–133
    competition from, 5, 66

Raubsville, 63, 64

Reber Brothers, 117

reclamation of coal, 110–112

recreational use of canals, 9, 36, 44, 45, 50–51, 57, 77, 83, 114

Reed, George, 109

Reed, Howard, 43

Reed, Joe, 84

Reed, William, 105

repair yards, 32

repairs. *See* maintenance

Rice, Keiser, 106

Riegelsville, 61–62

River Road, 58

round trip distances, 120, 142

rudder, 29

**S**

Saeger's Mill, 97

sale of water, 63, 75

Schimer Foundry, 86

Searfoss, George and John, 111

section boats, 12, 22, 48

Seitz's beer boat, 115, 134

Sheetz, Samuel, 31

Sherman, Ted, 130

Shipman, E.H., 19

shovel dredge, 37

Siegfried, 104, 106, 107

Sigafoos store, 53

silk industry, 95

Singley Store, 55

Singley, Joseph, 106

size of boats, 33, 54

size of locks, 5, 33, 54, 64, 69, 150

slackwater pools, 71, 74, 95, 99

Slate Dam, 108, 111

Snufftown, 65, 66

Solt, Harry E., 129

speed, 19, 130

steamboats, 94, 109, 110

stiff boats, 48

stone boats, 25, 38

stop gates, 62

stores, 47, 53, 55, 113

Stout, Bob, 83

strike, great, 146

Sundays on canal, 36, 44

supplies, 47, 82

swimming in canal, 83

Switchback Railroad, 142

Switzerland of America, 126

Swope family, 7

**T**

tainter gate, 54

Tattersall's coal yard, 22

taverns, 47. *See also* hotels

Taylorsville, 25

Thomas Iron Co., 103

Thomas, David, 102

tide lock, 9, 10

Tiffany, Louis, 6, 56, 124

Tinicum Creek aqueduct, 48

Tinsman family, 38

Tohickon Creek aqueduct, 44

toll collection, 10, 31

tonnage carried, 5, 12, 27, 66, 91, 141

towing post, 54, 85, 105

tractor hauling boat, 120

train hauling coal, 88, 114

transportation of anthracite, 91

Treasure Island, 45

Treichler's, 110, 112

Tuckerman, Walter, 6, 56, 124

Tullytown, 18, 20

**U**

Uhler, Peter, 52

Uhlertown, 52–53

Union Mills Paper Company, 26, 30

unloading boats, 22, 42, 93, 115

Upper Black Eddy, 55

Upper Grand Section, 144–153

**W**

Wahnetah, Hotel, 147

Walck, Fred, 64, 109

Walnutport, 113–115

warehouses, 73

water supply for canal, 26, 71, 137

water, sale of, 63, 75

weigh lock, 128–129

Weissport, 122–123
    boatyard, 122–123
    closing of, 108

Wells, Frank, 130

wheels, water, 26

White Haven, 151

White Haven dam, 152

Williamsport, 65, 66, 70

winter, 49, 55

wire rope, 5, 80, 143

Wood, Lovett, 96

work boats, 36, 46, 79, 99, 101, 107, 113, 117

work day for mules, 130

workers' housing, 80

**Y**

Yardley, 23–25

Young, John W.G., 45

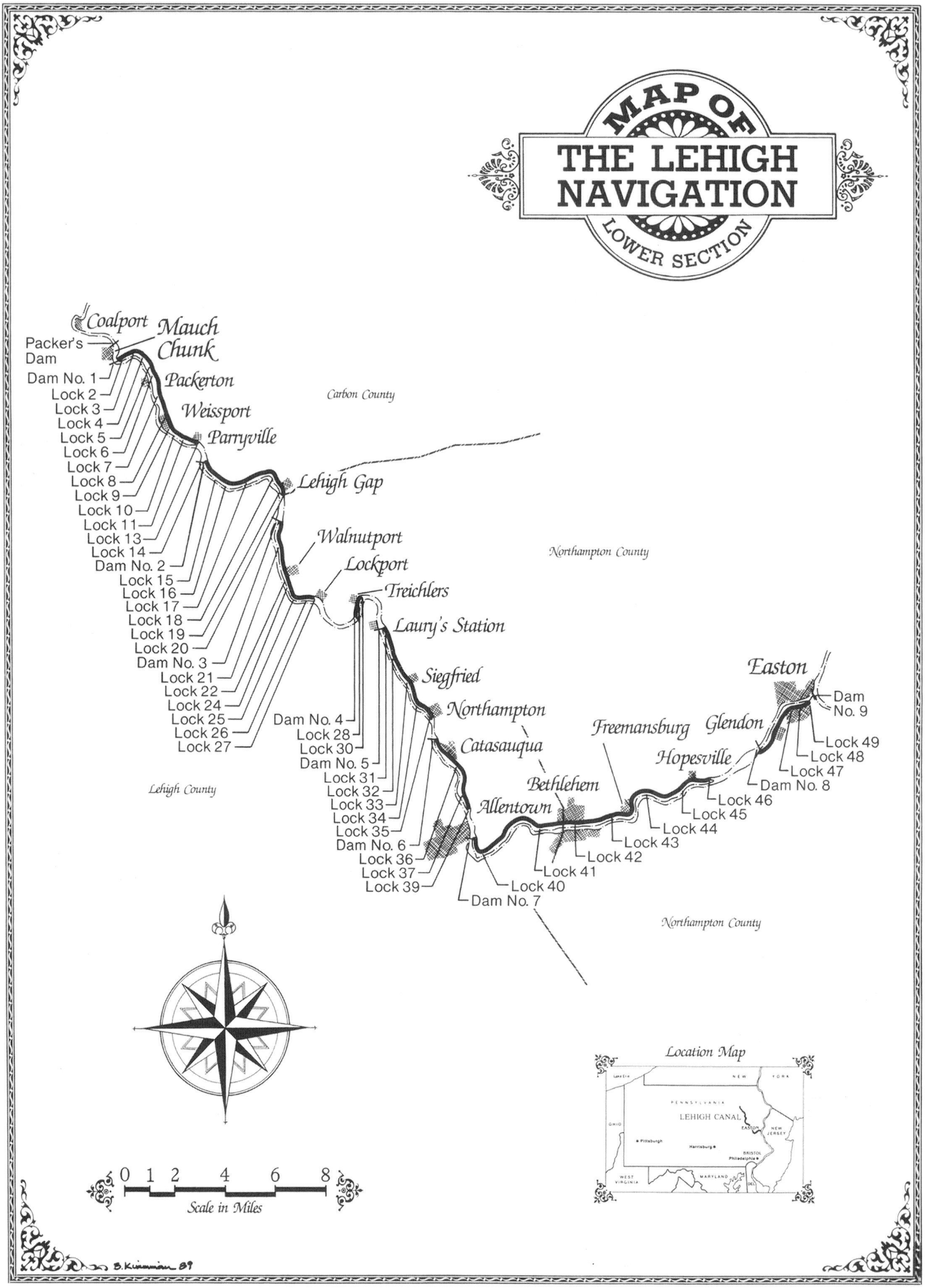

MAP OF
THE LEHIGH
NAVIGATION
LOWER SECTION

Carbon County

Northampton County

Lehigh County

Northampton County

Coalport
Mauch Chunk
Packer's Dam
Dam No. 1
Packerton
Lock 2
Lock 3
Weissport
Lock 4
Lock 5
Parryville
Lock 6
Lock 7
Lock 8
Lock 9
Lehigh Gap
Lock 10
Lock 11
Lock 13
Lock 14
Walnutport
Dam No. 2
Lockport
Lock 15
Lock 16
Treichlers
Lock 17
Lock 18
Laury's Station
Lock 19
Lock 20
Dam No. 3
Lock 21
Siegfried
Lock 22
Lock 24
Northampton
Lock 25
Lock 26
Lock 27
Dam No. 4
Catasauqua
Lock 28
Lock 30
Dam No. 5
Lock 31
Lock 32
Bethlehem
Lock 33
Allentown
Lock 34
Lock 35
Dam No. 6
Lock 36
Lock 37
Lock 39
Lock 40
Lock 41
Dam No. 7
Lock 42
Freemansburg
Glendon
Easton
Hopesville
Lock 43
Lock 44
Lock 45
Dam No. 8
Lock 46
Lock 47
Lock 48
Lock 49
Dam No. 9

Location Map

Lake Erie
NEW YORK
PENNSYLVANIA
OHIO
LEHIGH CANAL
EASTON
NEW JERSEY
Pittsburgh
Harrisburg
BRISTOL
Philadelphia
WEST VIRGINIA
MARYLAND
DEL.

0  1  2    4    6    8
Scale in Miles

B. Kummau 89